Wilhelm Hildebrand, John A. Roebling's Sons Company

The Underground Haulage of Coal by Wire Ropes

including the system of wire rope tramways as a means of transportation

for mining products

Wilhelm Hildebrand, John A. Roebling's Sons Company

The Underground Haulage of Coal by Wire Ropes
including the system of wire rope tramways as a means of transportation for mining products

ISBN/EAN: 9783337187248

Printed in Europe, USA, Canada, Australia, Japan

Cover: Foto ©berggeist007 / pixelio.de

More available books at **www.hansebooks.com**

BY WIRE ROPES.

INCLUDING THE SYSTEM OF

WIRE ROPE TRAMWAYS

AS A MEANS OF TRANSPORTATION FOR MINING PRODUCTS.

A PRACTICAL ESSAY,

WRITTEN FOR

JOHN A. ROEBLING'S SONS CO.

OF TRENTON, N. J.

BY WILHELM HILDENBRAND, C. E.

PRINTED BY THE W. S. SHARP PRINTING CO., TRENTON, N. J.

1884.

PREFACE.

In the present little volume we offer to our patrons, mining engineers, and to all interested in the transportation of coal and other mining products, a collection of such useful, practical and theoretical information as will be of assistance in planning appropriate machinery for hauling and hoisting.

We have refrained from filling the book with pictures and fancy sketches, and include only a series of carefully selected plans, representing the executed and approved conveying systems operated at the present day in the best American and English collieries.

It was originally our intention to confine this treatise solely to "underground" haulage, but while preparing the same we received so many inquiries about wire rope tramways that we considered the subject but incompletely treated without describing this method of "overground" transportation, which in many mining districts already plays an important part, and promises with every year to become more and more extended.

The general plans, showing the disposition of the systems, are invariably sketched without scale, as it would have been impossible to illustrate, in a correctly proportioned drawing, all the different parts.

The details, however, are drawn to a scale, and in most cases the dimensions are added in figures, justifying us to trust that in combination with the explanatory text the subject may be clear even to the non-professional reader.

Of the numerous wire rope planes constructed in England and Scotland, we selected, according to our judgment, some of the best systems, and included such details as differ from those used in our country, and which may be of interest.

The majority of examples, however, are taken from the Monongahela coal region, where we personally collected the general data and made sketches and measurements on the ground of all important details.

We are much indebted to the owners of the different establishments mentioned herein who kindly permitted our engineer to visit their works; also to the superintendents, engineers and pit bosses who courteously gave him ready and instructive information, enabling him to make an intelligent compilation and classification of the vast variety of arrangements. To all of these gentlemen we tender herewith our most sincere thanks.

We hope that this little book, which as far as we know treats the subject of haulage more comprehensively than any hitherto published, will be favorably received by mine operators. It will be highly gratifying to us if we should succeed in our object of contributing something to the general knowledge of underground and overground transportation, and of assisting parties in applying to the best advantage one or the other of the described systems.

JOHN A. ROEBLING'S SONS CO.

TRENTON, N. J., *June, 1884.*

THE UNDERGROUND HAULAGE OF COAL.

With the steadily increasing demand of coal for all purposes of industry, deep mining becomes more and more necessary. This is true not only in those regions where coal occurs at great depths, and must be reached by sinking deep shafts, but also in the more favored localities where the veins crop out at the hillside, and the miner, following the strata, is compelled to penetrate for miles into the heart of the mountain. In all cases it is a question of great importance how to convey the coal from the interior working "rooms" to the bottoms of the shafts, or directly to the surface, and from there to suitable shipping places, and it is probably not saying too much to assert that coal-mining, considered as an industrial and commercial success, at the present day is principally dependent upon the methods by which this is done. It is easy to understand that sinking numerous shafts in developing mining properties must be expensive and inconvenient, and that it is preferable to transport the coal underground, even great distances, to one centrally located shaft, if this can be done quickly and economically. This is fully demonstrated in the deep mines of England and the European Continent, where for the last twenty-five years the underground haulage of coal by machinery has superseded the older methods. The coal extending over a field of several square miles is now conveyed to the surface through a single deep shaft cheaper and in less time than formerly, where the coal could be mined near the surface, through a number of shallow shafts placed only a few hundred yards distant from each other. Moreover, the superior machinery for hauling, hoisting and pumping makes it possible to locate this shaft either in the deepest or any other part of the mine, wherever it is most advantageous for draining the water or landing the coal.

In the Monongahela and Ohio coal regions of this country the usual method of mining is by horizontal or slightly dipping "entries," and in the anthracite region by "slopes" and "gangways," through which the coal is brought to the surface without the necessity of vertical hoisting, but it frequently has to be transported long distances within the mine itself. The economy in the use of machinery is also well understood, and in many mines extensive appliances of machinery have been made. Wherever manual or animal labor for transporting coal is still employed, the mine owners contemplate replacing them by steam power, and it is a question of only a comparatively short time when every coal mine in the country will have efficient and improved mechanical arrangements for conveying coal from the interior to the "tipple," or place of shipping.

If we consider that as recently as seventy years ago, in England and Scotland, coal was carried to the surface by women, on their heads; that wheelbarrows or sledges, dragged by hand or by dogs, were used for a long time; that hoisting was done by horses in gins or by water-balance shaft; that even after the introduction of the iron rail until a recent date horses and mules were exclusively employed, all of which could transport only limited quantities of coal—we can better appreciate the immense advantages of modern progress in the perfection of machinery with which now more coal is brought to the surface of the earth in a day than half a century ago was brought in a year.

In the following treatise we intend to describe the methods employed in the best mines of this country, as well as of Europe, where coal is handled by means of wire rope haulage, which has been proved to be the cheapest and most convenient of any of the methods now in use. We will also give some practical hints and sufficient illustrations of detail construction to enable managers of mines to judiciously arrange their plant and to select that system which will be most suitable to their special circumstances.

The many methods applied in mining regions for transporting coal by means of wire rope, though varying from each other in detail, can be grouped in five distinct classes:

    I. The Self-Acting or Gravity Inclined Plane.
    II. The Simple Engine Plane.
    III. The Tail Rope System.
    IV. The Endless Rope System.
    V. The Cable Tramway.

I. THE SELF-ACTING INCLINED PLANE.

The motive power for the Self-Acting Inclined Plane is gravity; consequently this mode of transporting coal finds application only in places where the coal is conveyed from a higher to a lower point, and where the plane has sufficient grade for one or more loaded descending cars to raise the same number of empty cars to an upper level. Inside the pits such favorable circumstances are rarely met with, but they are of frequent occurrence where the coal is transported from the pit-mouth to the tipple. There is hardly a single mine in the Monongahela valley without a gravity inclined plane, varying in length from 700 to 1800 feet. The cars are raised and lowered by a wire rope winding and unwinding successively on a drum at the head of the plane. *Figs. 1* and *2, Plate 1,* on opposite page, show the general arrangements in plan and elevation. *Fig. 2A* represents a road with four rails, or a regular double track, the cars descending on one and ascending on the other. This is of course the safest, though most expensive, arrangement. There are no switches or crossings; it cannot get out of order, and will accommodate the transportation of unlimited quantities of coal.

Next in completeness is a road with three rails and a " parting " (*Fig. 2B*), an arrangement hardly inferior to the first, and considerably cheaper, saving one rail and not requiring so wide a road. The " parting " is a double track for a

The selecting Gravity inclined Plane

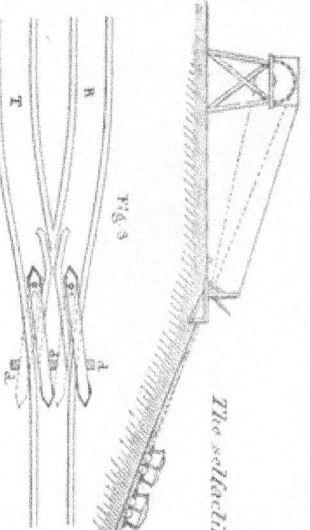

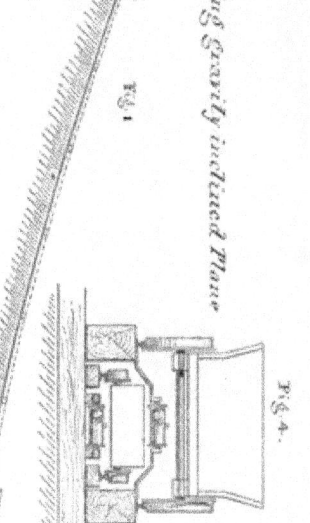

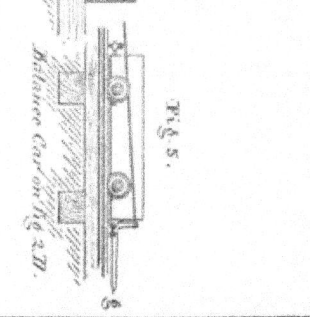

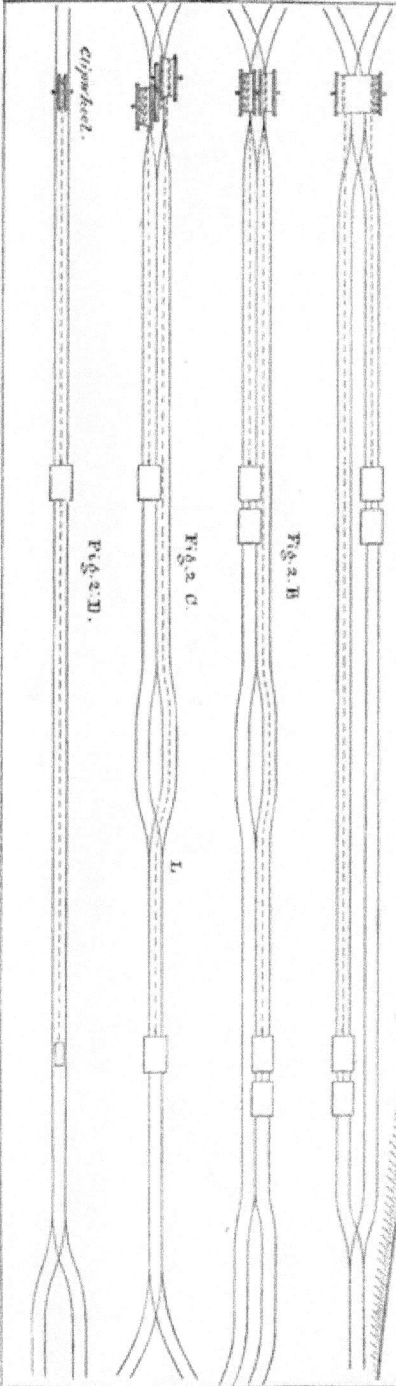

Fig 3

R

T

Fig 1

Fig. 4.

Fig. 5.

Selecting switch at L of fig 2 C.

Clipwheel.

Fig. 2 D.

Fig. 2 C.

L

Fig. 2 B

Fig. 2 A.

Balance Car on fig. 2 D.

short distance in the middle of the plane, where the cars meet and are allowed to pass each other.

Another still more economical plan (*Fig. 2C*) consists, like the former, of a road with three rails from the top of the plane to the parting, but with two rails only below the same. To operate this plane it requires a self-acting switch at the lower end of the parting (*Fig. 3*). Two pieces of timber, pointed and iron-bound at the ends, are movable over the rails around the pivots (c). In the position, as drawn, an empty car going up will take the track (T), while the loaded car coming down the track (R) will shift the timbers to the position indicated in dotted lines. At the next trip the empty car will move on track (R), and the loaded car, this time descending on track (T), will shift the timbers again to their original position, the same play being repeated at every alternate trip. Some wooden or iron blocks (d) let in the ground or fastened to the ties, limit the motion of the timbers, which at the same time serve as guard-rails and guide the wheels on the respective tracks. The switch is simple and will not easily get out of order, and as long as this is the case this plane will do the same service as one with a double track.

In narrow pits, or wherever space must be economized, an arrangement as illustrated in *Fig. 2D* and *Fig. 4*, may be recommended. It consists of a double track, one inside the other; the outer for the passage of the pit cars, the inner, constructed of lighter rails, for an extra balance car. The latter is a hollow iron box, on small wheels, and heavy enough to pull an empty car up the plane, and also light enough to be raised by a descending loaded car. At the Westmoreland colliery a plane of this construction, 500 feet long, with a grade of one in three, is in operation for lowering material into the mine. The balance car (*Fig. 5*) is 4x2 feet, by ten inches high, resting on six-inch wheels. It is provided with a safety bar (g) attached by a pin (o) to the front of the car. The wire rope, passing through the box, is secured to a short arm of the bar, and in case of breaking or getting slack the latter will drop and arrest the motion of the car. The wooden rollers (p) serve for the support of the rope. The working capacity of such a plane, where coal can be lowered only at every other trip, is just half of that of the three former arrangements.

At the head of every gravity plane there is a drum, which is generally constructed of wood, having a diameter of seven to ten feet. It is placed high enough to allow men and cars to pass under it. It is enclosed by a small shed, technically called the "check-house," by which name it is known throughout the mine regions. Loaded cars, coming from the pit, are either singly or in sets of two or three switched on the track of the inclined plane, and their speed in descending is "checked" or regulated by a brake on the drum. In place of one large drum, as shown in *Fig. 2A*, two smaller ones, keyed to the same shaft, are frequently used (*Fig. 2B*). The axis of the drum is horizontal. Two wire ropes are fastened with their ends to the circumference of the drum, one leading from the under, the other from the upper side of it, so that one must wind up while the other unwinds. Sometimes the two drums are on different shafts, which, by means of a gearing of two (36-inch) cogwheels, revolve in opposite directions. This makes it possible to lead both ropes from the under side of the drums, which is preferable on account of avoiding any tendency to lift the cars when they are near or at the top of the plane. Instead of using two ropes, each of the length of the plane, it has been attempted to obtain the same result with a single length of rope. It is done by giving it three to four turns on the drum, to prevent slipping, and passing the ends as before, one from the under the other from the upper side of the drum. For planes in constant use this method is not advisable on account of the unavoidable sideward sliding of the rope on the drum, which wears it out more rapidly. The use of a "grip-wheel" is better for this purpose. It consists of two rims bolted together. Numerous hinged

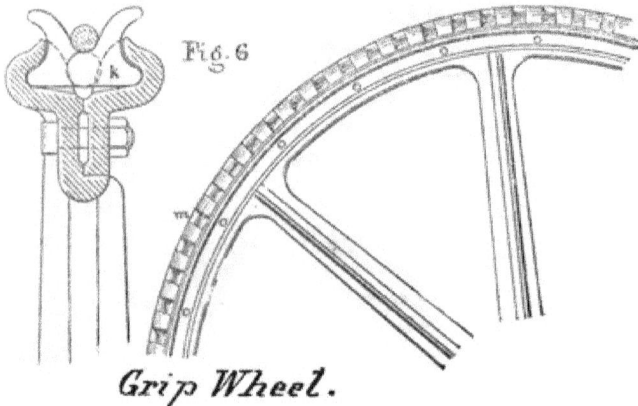

Fig. 6

Grip Wheel.

jaws (k) are fitted into the square holes (m) in the circumference of the rim. The lower side of the jaws being straight, while that of the holes is slanting, the pressure of the rope will cause the upper part of the jaws to close, gripping the rope firmly and preventing its slipping. A grip-wheel occupies very little space, and has also the advantage of requiring only one length of rope, but the first method, with a drum and a double rope, recommends itself by its simplicity of construction and certainty of action, and in most cases is preferable.

The braking apparatus, as used at the Imperial inclined plane, is illustrated in *Fig. 7, Plate 2*. There are two drums on the same shaft; each drum is provided with a brake-band (b), fastened on opposite sides, with one end to the supporting frame, with the other to the circumference of a small disk (d), in such a way that both bands will tighten if the disks are turned in the direction of the arrows. The tightening is effected by a man on the platform at the brink of the plane in pressing on the lever (l), a wire leading from it to the two levers (p), which are attached to the axles of the disks.

Fig. 8 represents the typical form of the switch used in the check-house at the head of the plane. It is taken from the D. Steen mine, and consists of four movable tongues (a, b, c, d), which, according to their positions, turn the loaded cars alternately on tracks (L) and (R), while all empty cars are returned on track (M). In the position as drawn a loaded car will pass from track (N) to

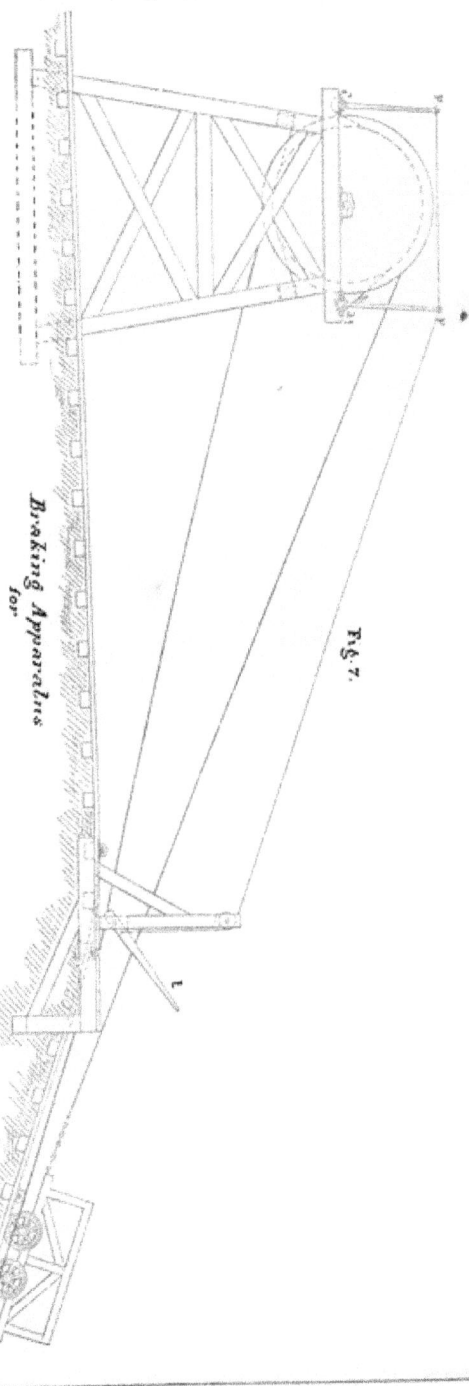

Fig. 7.

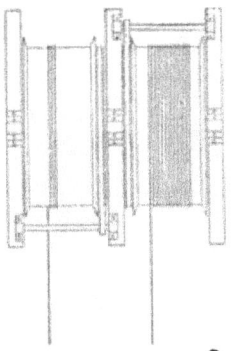

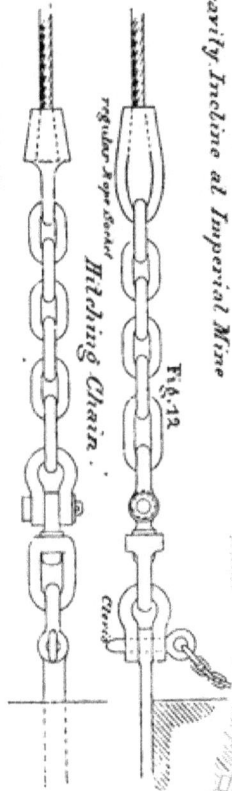

Braking Apparatus
for
Gravity Incline at Imperial Mine

Fig. 12.

Hitching Chain.

regular rope socket

Clevis

Pl. 2

(L); the empty car returning from the direction (R), will open the tongue (d) and **close** (c); (a) is closed by the switch-tender, and (b) is constantly kept partially open by inserting a piece of wood between the tongue and the fixed rail. By this means there is enough spring imparted to the tongue (b) to allow the

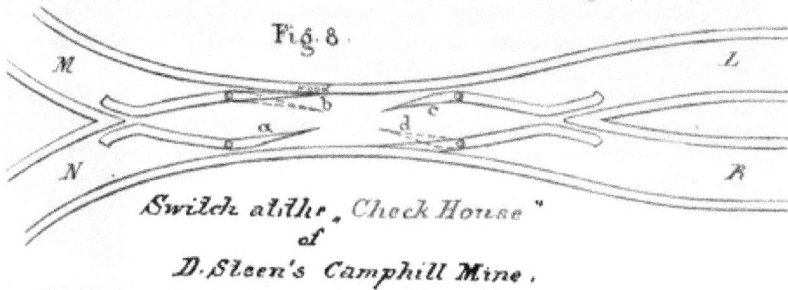

Fig. 8

Switch at the „Check House"
of
D. Sleen's Camphill Mine.

next loaded car to pass over to track (R), while the returning empty one will place the **tongues** in their original position, **and will again turn** on track (M). The switch **is, therefore,** self-acting, with the **exception that the tongue** (a) must be closed by the switch-tender for every returning empty car.

Safety Lock
on Gray and Bell's gravity incline

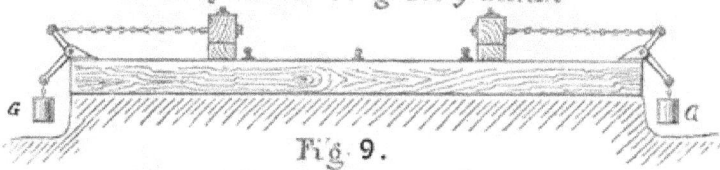

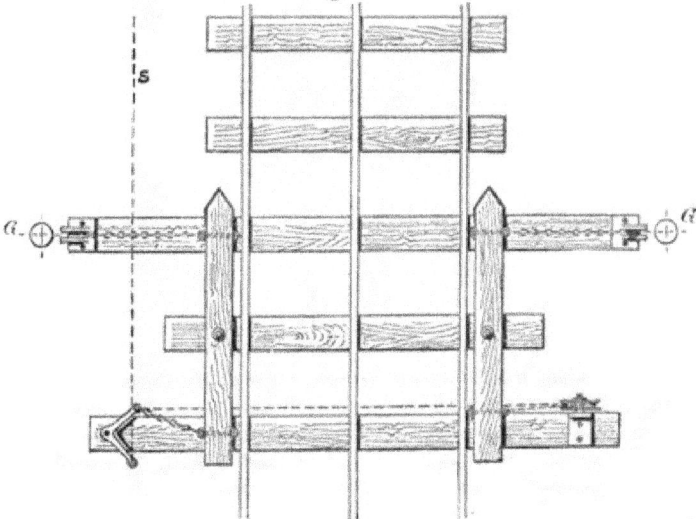

Fig. 9.

Every gravity inclined plane is provided with safety arrangements for guarding against damage to life and property in case of breakage of the rope. There are a variety of devices at different mines, differing in detail construction, but in all of them only two principles are represented: either to stop the cars in their downward motion, or to switch them sideways and throw them off the plane. An arrangement for stopping the cars, called the "dead-lock," which is in use at the Gray & Bell mine, is illustrated in *Fig. 9*.

Two pieces of timber, pointed and iron-bound at the upper ends, and movable around central pivots, can be placed over the rails by pulling the wire (s) which leads to the check-house, and which by means of a bell-crank is connected with the lower ends of the timbers. When the tension on this wire is released, the timbers are drawn into their original position by the weights (G) which are suspended from bell-cranks at the end of wires attached to the front part of the timbers.

Another arrangement intended to throw the cars off the track, and in use at the Imperial mine, is the following: A regular switch (*Fig. 10*) is placed in

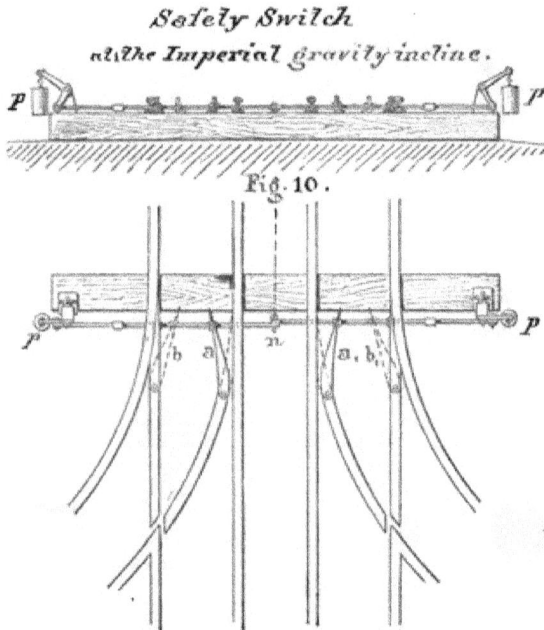

Safety Switch at the Imperial gravity incline.

Fig. 10.

a position to cause descending cars to run off the track in case of breakage of the rope. This is effected by the weights (P) hanging at the ends of double-armed levers, which have a tendency to press the switch-rods towards the centre of the road, closing the tongues (a) and (a_1), and opening (b) and (b_1). The switch-rods are held in position by a pin (n), and the weights (P) come only in action when the pin is pulled out by means of a wire leading to the check-house or to the tipple, according to the location of the switch—whether nearest to the top or to the bottom of the plane.

At the Walton mines a heavy timber, called the "dead-fall," moving in an upright frame, is made to fall over the rails. At Brown's mine, iron blocks—"dead-locks"—are pushed in the way of the car wheels, and many other devices of a similar character are employed at other mines.

Important parts of every gravity plane, or of any other road worked by wire rope, are the supporting rollers, which are placed between the track at regular distances. Without them the rope would drag on the ground and wear out rapidly. They are generally of wood, five to six inches in diameter and eighteen to twenty-four inches long, with $\frac{3}{4}$ to $\frac{1}{2}$-inch iron axles. Frequently the bearings

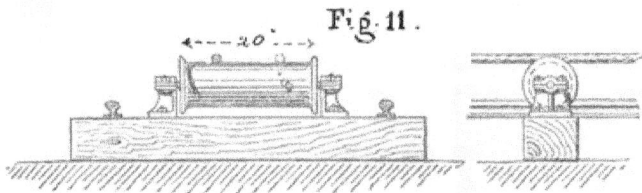

Fig. 11.

are also of wood, but on the better constructed roads they are of iron, the boxes being hinged so that the rollers can easily be changed. The wood preferred for rollers in the Monongahela region is sugar-maple or gum-wood, and the price per piece is twenty-five to thirty cents, while the first cost with axles and bearings is about $1.25. If partially worn, the rollers are turned around or the bearings shifted sideways, by which means the rollers are made to last from six months to two years, according to the quality of the wood and the condition of the road. The distance between the rollers varies from fifteen to thirty feet—steeper planes requiring less rollers than those with easy grades. Considering only the reduction of friction and what is best for the preservation of the rope, a general rule may be given to use rollers of the greatest possible diameter, and to place them as close as economy will permit.

Fig. 13

Extra slender Socket.

Fig. 14

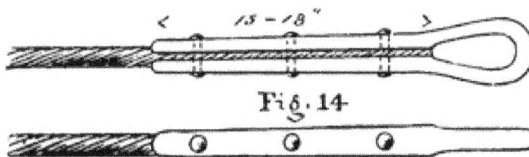

The "Gooseneck" Socket.

Wire-rope fastenings, as used in the mine regions, consist either of a regular conical socket (*Fig. 12, Plate 2*), secured to the rope by tapering pins driven between the opened wires, or more frequently of an especially long, thin socket, called a " dead-eye " (*Fig. 13*), to which the rope is fastened simply by turning over the strands or sometimes by opening the wires inside the socket and pouring in melted lead.

The most usual connection, however, is the so-called " goose-neck " (*Fig. 14*). It consists of a pair of trough-shaped tongs, bent to a loop and riveted to the rope by three or four rivets. The rivets are driven cold and the holes are made by inserting a sharp pin and pushing the wires apart without injuring them. It is a good plan to heat the end of the rope for a length of about one inch, and weld the wires together to prevent them from untwisting. Of these three fastenings, the first is the only one that gives the same strength as the rope; but the others, being neater and more compact, slide easily over the rollers, and in most cases are sufficiently strong, as the rope is generally strained only from one-fifth to one-seventh of its ultimate strength. The rope socket can be directly attached to the " pulling bar " of the car by a clevis, or it may first be connected to a chain five to ten feet long, with a swivel, the end of the chain being attached to the car (*Fig. 12, Plate 2*). The latter plan is preferable, because the greater flexibility of the chain avoids the cross-strains in the rope caused by vibrations. The wires near the socket generally break long before the rope is worn out, and it is therefore good practice to cut off the socket once a year and place it a few feet farther back.

It will be of interest to know the smallest angle of inclination at which a plane can be made self-acting. The limit will be when the motive and resisting forces balance each other. The motive forces are the gravity of the loaded car and of the descending rope. The resisting forces consist of the weight of the empty car and ascending rope, of the rolling and axle friction of the cars, and of the axle friction of the supporting rollers. The friction of the drum, stiffness of rope and resistance of air may be neglected. A general rule cannot be given, because a change in the length of the plane or in the weight of the cars changes the proportion of the forces; also because the co-efficient of friction, depending on the condition of the road, construction of the cars, &c., is a very uncertain factor.

The subjoined table has been calculated upon the assumption of ordinary circumstances for working a plane with a ⅝-inch steel rope and lowering from one to four such pit cars as are used in the Monongahela mine region. The last column gives the strains on the rope in each case:

Number of Cars.	Weight of descending Cars.	Weight of ascending Cars.	Length of Plane.	Rise in 100 feet necessary to make the plane self-acting.	Strain in the Rope.
	Pounds.	Pounds.	Feet.	Feet.	Pounds.
1	4,000	1,400	500	5.5	135
"	"	"	1,000	6.3	174
"	"	"	1,500	8.2	240
"	"	"	2,000	10.0	318
2	8,000	2,800	500	5.0	231
"	"	"	1,000	5.5	270
"	"	"	1,500	6.1	314
"	"	"	2,000	6.7	364
3	12,000	4,200	500	4.9	314
"	"	"	1,000	5.2	367
"	"	"	1,500	5.5	404
"	"	"	2,000	5.9	449
4	16,000	5,600	500	4.8	412
"	"	"	1,000	5.0	463
"	"	"	1,500	5.2	497
"	"	"	2,000	5.5	539

At the inclination, given in the fifth column, the cars will descend with uniform velocity equal to an initial velocity, which may be given to them either by a push or by a "knuckle" (a short stretch of steeper inclination at the head of the plane). On a smaller inclination the cars will not be able to descend, except when, under very favorable circumstances, the friction should be much smaller than assumed in the above table. If the rise is greater, the brake must be applied to check the accelerated motion which the cars would otherwise assume.

A gravity inclined plane should be slightly concave, steeper at the top than at the bottom. The maximum deflection of the curve should be at an inclination of forty-five degrees, and diminish for smaller as well as for steeper inclinations.*

---

* The theoretically correct curve is the arc of a cycloid. This curve has the property that a body falling from one point of it to another will reach the lower point in less time than on any other curve or straight line, and that the lowest point or vertex will always be reached in the same time, no matter where the starting point has been. These two qualities have given to this curve the surnames, "Brachystochrone" and "Tautochrone"—that is, "curve of quickest descent," and "curve of equal-time descent." It possesses another quality of great practical value, lately demonstrated by Julius Ritter von Hauer (see Berg and Huettenmaennische Zeitung, 1884), that it equalizes the weight of the two ropes at every point, so that the resistance remains always the same, and the braking power can be applied with equal force during the whole time of the descent, while on a straight plane the necessary braking force changes considerably.

We subjoin another table, giving the maximum strain on a wire rope working gravity inclines of different lengths and with grades of from five to forty-five degrees. It is supposed that starting and stopping is done gradually and not with a jerk. A safety factor of six has been assumed in the determination of the rope diameter:

Angle of Inclination of Plane	Rise in 100 feet.	Number of Descending Cars.	Weight of Descending Cars.	Weight of Ascending Cars.	Length of Plane. 500 Feet. Strain in Rope.	500 Feet. Diam. of Steel Rope.	1,000 Feet. Strain in Rope.	1,000 Feet. Diam. of Steel Rope.	1,500 Feet. Strain in Rope.	1,500 Feet. Diam. of Steel Rope.	2,000 Feet. Strain in Rope.	2,000 Feet. Diam. of Steel Rope.
Degrees.	*Feet.*		*Pounds.*	*Pounds.*	*Pounds.*	*In.*	*Pounds.*	*In.*	*Pounds.*	*In.*	*Pounds.*	*In.*
5°	8.7	1	4,000	1,400	200	[illegible]	320	[illegible]	340	[illegible]	360	[illegible]
"	"	2	8,000	2,800	580	[illegible]	600	[illegible]	620	[illegible]	640	[illegible]
"	"	3	12,000	4,200	865	[illegible]	885	[illegible]	905	[illegible]	925	[illegible]
"	"	4	16,000	5,600	1,146	[illegible]	1,166	[illegible]	1,186	[illegible]	1,206	[illegible]
10°	17.6	1	4,000	1,400	673	[illegible]	718	[illegible]	763	[illegible]	808	[illegible]
"	"	2	8,000	2,800	1,302	[illegible]	1,346	[illegible]	1,390	[illegible]	1,434	[illegible]
"	"	3	12,000	4,200	1,930	[illegible]	1,975	[illegible]	2,020	[illegible]	2,065	[illegible]
"	"	4	16,000	5,600	2,558	[illegible]	2,603	[illegible]	2,648	[illegible]	2,693	[illegible]
15°	26.8	1	4,000	1,400	1,038	[illegible]	1,108	[illegible]	1,178	[illegible]	1,248	[illegible]
"	"	2	8,000	2,800	2,009	[illegible]	2,079	[illegible]	2,149	[illegible]	2,219	[illegible]
"	"	3	12,000	4,200	2,980	[illegible]	3,050	[illegible]	3,120	[illegible]	3,190	[illegible]
"	"	4	16,000	5,600	3,965	[illegible]	4,050	[illegible]	4,135	[illegible]	4,220	[illegible]
20°	36.4	1	4,000	1,400	1,396	[illegible]	1,489	[illegible]	1,582	[illegible]	1,675	[illegible]
"	"	2	8,000	2,800	2,703	[illegible]	2,796	[illegible]	2,889	[illegible]	2,982	[illegible]
"	"	3	12,000	4,200	4,028	[illegible]	4,142	[illegible]	4,256	[illegible]	4,370	[illegible]
"	"	4	16,000	5,600	5,364	[illegible]	5,508	[illegible]	5,652	[illegible]	5,796	[illegible]
25°	46.6	1	4,000	1,400	1,746	[illegible]	1,862	[illegible]	1,978	[illegible]	2,094	[illegible]
"	"	2	8,000	2,800	3,402	[illegible]	3,544	[illegible]	3,686	[illegible]	3,828	[illegible]
"	"	3	12,000	4,200	5,069	[illegible]	5,250	[illegible]	5,431	[illegible]	5,612	[illegible]
"	"	4	16,000	5,600	6,747	[illegible]	6,975	[illegible]	7,203	[illegible]	7,431	[illegible]
30°	57.7	1	4,000	1,400	2,080	[illegible]	2,219	[illegible]	2,358	[illegible]	2,497	[illegible]
"	"	2	8,000	2,800	4,054	[illegible]	4,224	[illegible]	4,394	[illegible]	4,564	[illegible]
"	"	3	12,000	4,200	6,098	[illegible]	6,369	[illegible]	6,640	[illegible]	6,912	[illegible]
"	"	4	16,000	5,600	8,131	1	8,495	1	8,859	1	9,223	1
35°	70.0	1	4,000	1,400	2,399	[illegible]	2,558	[illegible]	2,717	[illegible]	2,876	[illegible]
"	"	2	8,000	2,800	4,727	[illegible]	4,975	[illegible]	5,223	[illegible]	5,471	[illegible]
"	"	3	12,000	4,200	7,031	[illegible]	7,344	[illegible]	7,657	[illegible]	7,970	[illegible]
"	"	4	16,000	5,600	9,377	1	9,797	1	10,217	1	10,637	1
40°	83.9	1	4,000	1,400	2,699	[illegible]	2,879	[illegible]	3,059	[illegible]	3,239	[illegible]
"	"	2	8,000	2,800	5,392	[illegible]	5,744	[illegible]	6,096	[illegible]	6,448	[illegible]
"	"	3	12,000	4,200	8,031	1	8,503	1	8,975	1	9,447	1
"	"	4	16,000	5,600	10,652	1⅛	11,224	1⅛	11,796	1⅛	12,368	1⅛
45°	100.0	1	4,000	1,400	3,024	[illegible]	3,266	[illegible]	3,510	[illegible]	3,754	[illegible]
"	"	2	8,000	2,800	5,950	[illegible]	6,339	[illegible]	6,728	[illegible]	7,117	[illegible]
"	"	3	12,000	4,200	8,864	1	9,385	1	9,906	1	10,427	1
"	"	4	16,000	5,600	11,916	1¼	12,708	1¼	13,500	1¼	14,292	1¼

II. THE SIMPLE ENGINE PLANE.

The name "engine plane" has been given to an inclined plane on which a load is raised or lowered by means of a single wire rope and stationary steam engine. It is a cheap and simple method of conveying coal underground, and therefore is applied wherever circumstances permit it. It requires only a single track, a rope of the length of the plane, and the power of the engine only half the time. The road may be curved and may have variable grades, provided the fall is in one direction and of sufficient inclination to enable a full or empty set of cars to descend by force of their own gravity, dragging the rope after them. The smallest grade at which this is possible depends on the length and condition of the road, as well as on the weight of the cars and the rope. Under ordinary conditions, such as prevail in the Pennsylvania mine region, a train of twenty-five to thirty loaded cars will descend, with reasonable velocity, a straight plane 5000 feet long, on a grade of $1\frac{3}{4}$ feet in 100, while it would appear that $2\frac{1}{2}$ feet in 100 is necessary for the same number of empty cars. English authorities on this subject limit the grade to $3\frac{1}{2}$ feet in 100 for satisfactorily working an engine plane, but the English "tubs" compare unfavorably with the American "pit cars," requiring heavier grades to overcome the greater friction. It has been demonstrated in the Monongahela valley that engine planes, even with lighter grades than those mentioned above, work successfully, but it would not be safe to accept it as a rule. For roads longer than 5000 feet, or when containing sharp curves, the grade should be correspondingly larger.

A description of a few actually executed engine planes will give a clearer understanding of how they are operated under different circumstances, and what necessities of detail are advisable.

Hardley & Marshall's mine offers an example of an engine plane of the simplest kind for conveying the coal from a certain point inside the mine to the "tipple."[*] The road is 4600 feet long, with a total fall of 80 feet, commencing with three per cent. for the first 200 feet, then two per cent. for the next 1000 feet, and one and one-half per cent. for the remaining part; average grade, $1\frac{3}{4}$ feet in 100. The load to be raised consists of twenty-five to thirty cars of 4400 pounds each, pulled by a $\frac{7}{8}$-inch steel rope of seven wires to the strand. A round trip occupies nine minutes, and twenty trips are made per day, giving a daily output of nine hundred to nine hundred and fifty tons of coal. The returning train descends the plane with a velocity of fifteen miles per hour. An empty pit car weighs 1200 pounds, and its dimensions are six feet long, four feet wide and twenty inches high, with wheels eighteen inches in diameter fixed to the axles. The gauge of the track is two feet, and the rails weigh twenty-five pounds per yard. The rope drum, standing on top of the plane, has a diameter of five feet, and is driven by a single-acting engine with a fourteen-inch cylinder, having a thirty-inch stroke, a three-ton fly-wheel of nine feet diameter, and

---

[*] This name has been given in the Monongahela coal region to an elevated platform, some thirty or forty feet above the railroad or river, from which the coal is dumped over a screen into railroad cars or boats.

Pl. 3

Fig. 16.

Hoisting Engine
working the Shaft of the Standard Mine.

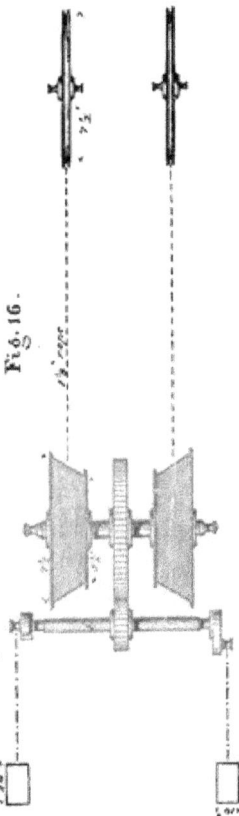

Fig. 15.

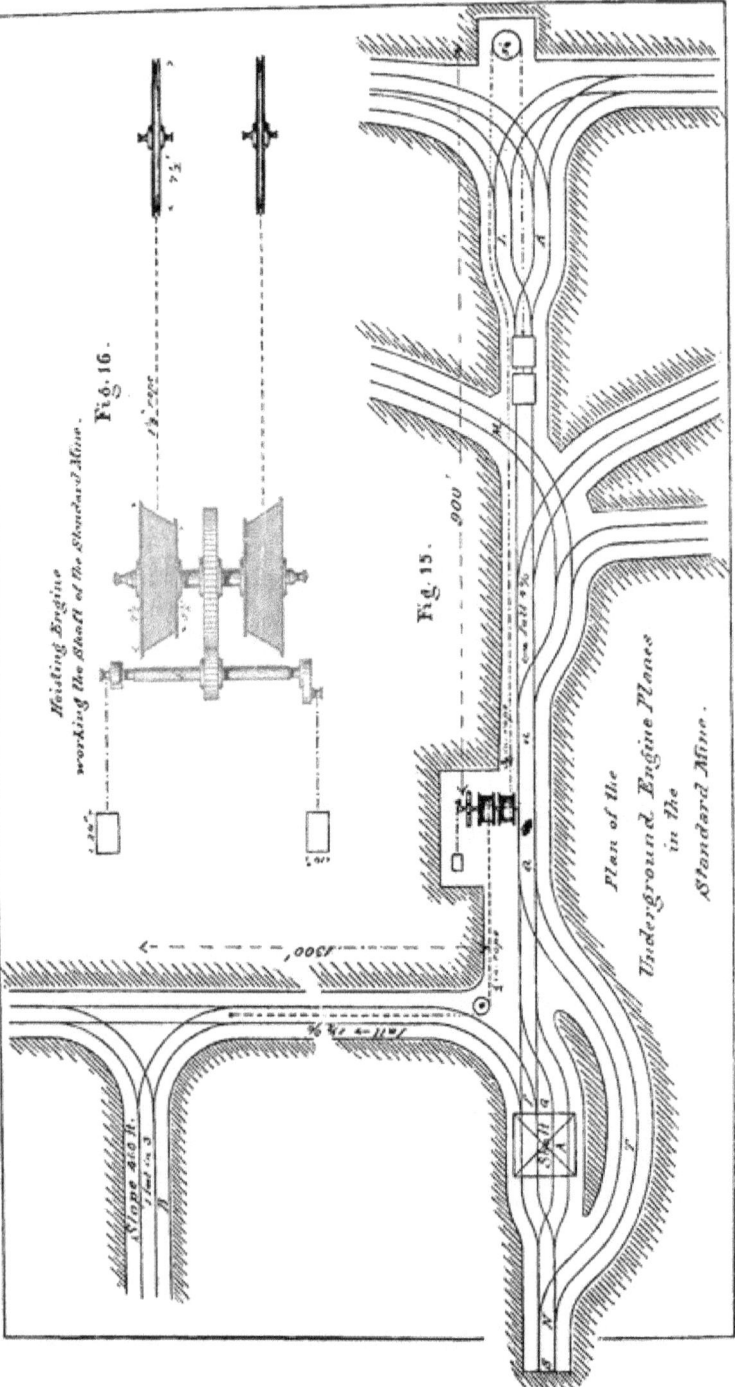

Plan of the
Underground Engine Planes
in the
Standard Mine.

gearing in the proportion of 1 to 2½. A wire rope, working this plane, lasts about four years, and the supporting rollers, which are placed twenty-seven feet apart, last about two years.

Sometimes it is more convenient to locate the engine at the bottom of the incline. In this case the rope, which must have twice the length of the road, is led from the drum around a sheave called the return-wheel, at the head of the plane.

An example of this kind, illustrated in *Fig. 15, Plate 3*, is taken from the Standard Mine and Coke Works. As will be noticed from the general ground plan, the coal (amounting to 1500 tons per day) is brought to the surface partially through a shaft (*A*) and partially on a slope (*B*). The latter is a simple engine plane, 400 feet long, rising one foot in three, and lifting three loaded cars with a ½-inch steel rope of nineteen wires to the strand. The shaft, containing a double "cage," has the capacity of raising, in eighteen seconds, two full cars to a height of 240 feet, while in the same time two empty ones are lowered. It is worked by an eighty-horse power engine, shown in its general disposition in *Fig. 16*. Near the bottom of the shaft there is a small engine, provided with steam from the boilers on top, which works two underground engine planes. One, 1300 feet long and rising 1½ feet in 100, leads from the slope to the shaft. It is worked by a ½-inch steel rope, but is used only occasionally, when the slope cannot accommodate all the coal collected at its foot. The other and principal plane of the description above mentioned, leads from the shaft to the parting, a distance of 900 feet, with an average rising grade of four per cent. It is worked by a ¾ inch iron rope, with nineteen wires to the strand, which, after fifteen months' service, did not show any appreciable wear. The rope is supported every thirty feet by a wooden roller, the "empty" side of the rope running outside the track along the wall of the entry, and the "pulling" rope in the centre of the track. Before passing around the return-wheel it is led under the rails in order to clear the crossings at the parting, the same being the case for the empty rope at point (*M*). The following is the mode of operation: A set of empty cars, standing on track (*a-a*), is attached to the end of the rope and taken up the incline, a man riding on the first car. At point (*M*), called the "false station," about in the middle of the plane, the engineer stops, and from three to five cars are detached and switched into one of the side entries. A bell, rung by the train-rider, signals the engineer to start again, and the remaining fifteen cars are taken to the regular parting on track (*L*), the other track (*R*) being occupied by a set of full cars ready to descend. As all entries in this mine have considerable grade, each car, for the convenience of the miners and mule-drivers, is provided with a hand-brake after the pattern shown in *Fig. 17, Plate 6*. These brakes, and in addition to them a wooden block over the rail called the "lock," prevent the train from starting prematurely. After the rope has been changed from the front of the empty to the rear of the full set, the train-rider gives the signal to the engineer to take up the slack of the rope, clearing at the same time the "lock," which is turned in the position shown in the sketch (*Fig. 18*). If the train does not start immediately a little jerk on the rope will do it, the

18

train-rider taking his position on the last car, standing on the rope or coupling-chain. The downward speed is regulated by the engineer with a brake on the

Fig 18.

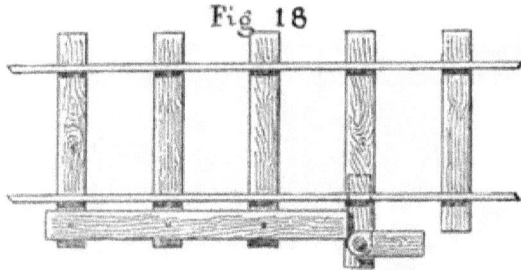

„Lock" in Standard Mine.

drum. To provide against accidents, in case the rope should break, a safety arrangement, called the "dead-latch" (see *Fig. 19*), near the foot of the incline, is

Fig 19.

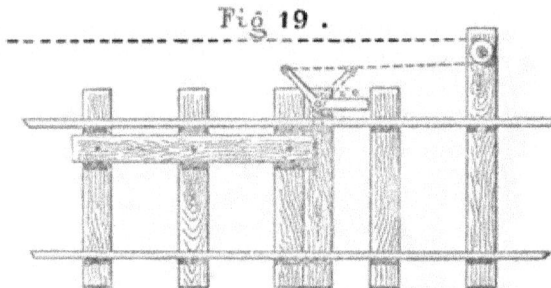

Safety „Latch" in Standard Mine

operated by the engineer, to whom the train-rider gives a bell signal. During the time occupied at the upper parting with changing the rope and getting ready for descending, three full cars are "braked" down from the intermediate station and the empty ones are returned by the next trip going up. From the foot of the inclined plane, the loaded cars, in sets of two, are pushed by hand on the cage of the shaft, alternately on tracks (*P*) and (*Q*), pushing at the same time the empty cars down to point (*N*) and up to (*S*), where they rebound and descend on track (*T*), which leads to the starting point (*a-a*) of the ascending set.

One of the oldest plants in this country for conveying coal by means of wire rope is at the mine of W. H. Brown, on the Youghiogheny river. It consists (*Fig. 20, Plate 4*) of two engine planes, one of one mile in length and a rise of 3¾ feet in 100, for transporting the coal from the parting within the pit to the pit mouth, the other for lowering it from the pit mouth to the tipple. The latter plane is one and one-eighth miles long, falling 7.6 feet in 100, and has several sharp curves. Each plane is worked by a separate engine, located on the crest of the inclines, near the pit mouth. A train of forty or fifty loaded cars,

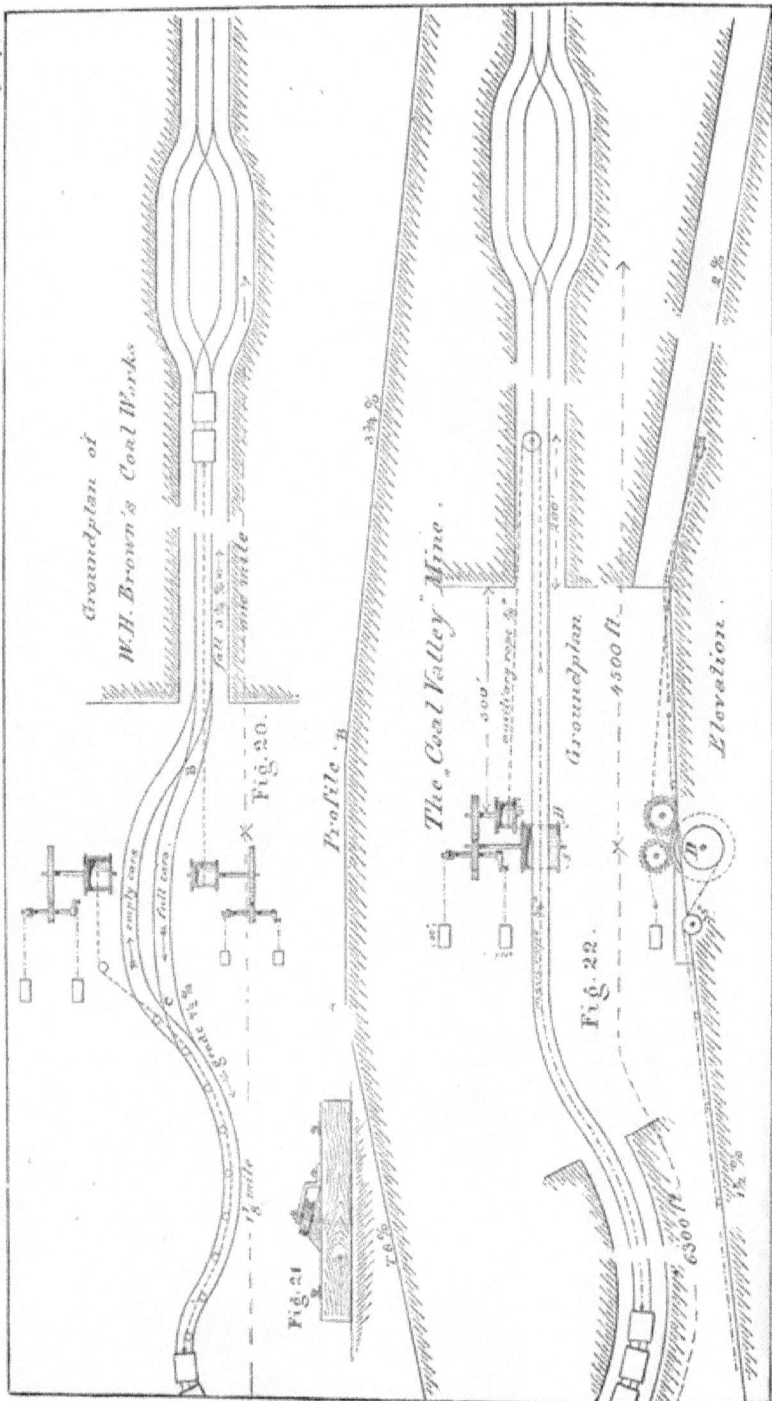

Pl. 4.

Groundplan of
W.H. Brown's Coal Works.

Fig. 20.

fall 3½ %

empty cars

full cars

grade 3½ %

1⅛ mile

Fig. 21.

7 6 %

Profile.

3½ %

The "Coal Valley" Mine.

500'

auxiliary rope ¾"

1000'

Fig. 22.

Groundplan.

4500 ft

6300 ft

1 ½ %

2 %

Elevation.

made up at the inner parting, is raised at the rate of twelve miles per hour to point (*B*), where the rope is "knocked off," the cars descending by their acquired momentum to point (*C*), on the left-hand track. Here the rope of the other engine is hitched to the tail-end of the train, which descends by force of its own gravity to the tipple. Now this rope is changed to the first car of an empty train and knocked off again, when the latter arrives at (*C*). Taking this time the right-hand track its momentum will carry it to (*B*), and from there will run alone into the mine, dragging the pit rope after it. The service is of course arranged so that both engines raise or lower at the same time, and each rope receives and despatches without delay the set just raised by the other rope. The first pit rope, which was of iron and of $\frac{7}{8}$-inch diameter, lasted six years, and the present one, also an iron rope, of $1\frac{1}{8}$-inch diameter, with seven wires to the strand, is still in good condition after six years' service. For the outside plane a $\frac{7}{8}$-inch steel rope has been in use for eleven years, and promises to last a few years longer. The durability of these ropes is perhaps partially due to the care with which they are treated ; every three months they receive a coating of pine tar, and all rollers are oiled every other day. The latter have a diameter of eight inches, and are placed twenty feet apart. They last from twelve to eighteen months. For the curves, cast-iron sheaves of 12-inch diameter are used, set between the two rails at a slight angle, as shown in *Fig. 21.* In the sharpest curves they must be replaced every six months, but in the easier ones they last one year or longer. About twenty trips are made per day, and the output is 19,000 bushels (at 76 pounds) of clear coal, and 4000 bushels of "slack." This name is given to the small pieces falling through a screen with $1\frac{1}{2}$-inch meshes, over which the coal is dumped into the railroad cars or boats.

The rope attachment consists of a goose-neck socket, fourteen inches long, riveted with three $\frac{7}{8}$-inch rivets, and coupled to a chain ten feet long, with swivel and clevis at the other end. In descending the outer incline, the coupling-chain is led under the last car and hitched to the pulling-bar of next to the last car, for the purpose of holding it down and making the rope drop into the guiding-sheaves at the curves.

The car-coupling is a rigid coupling, consisting of a clinch-hook, eighteen inches long by one and three-eighth inches square. In addition to it there are two safety-chains (see *Fig. 17*[b], *Plate 6*). These safety-chains form the only coupling for the last car on the train descending on the outer incline.

A parallel case to the plant just described is afforded at the mine of Foster, Clark & Wood, where two similarly situated engine planes are worked by a single engine and a single rope (*Fig. 22, Plate 4*). To accomplish this, the drum (*D*) is placed in the line of and under the main track, so that the cars can pass over it. A full train of forty-six cars starts from the parting within the pit, and is pulled up grade for a distance of 4500 feet, until the first car reaches the drum. Here the rope is knocked off, carried by hand over the drum and attached again to the rear car of the train, which in the meantime has traveled the distance of its own length by its acquired momentum. Now the drum is put out of gear and the cars descend the outer incline by force of gravity. The latter plane has a length of 6300 feet and falls 1½ feet in 100, while the pit plane has a rise of 2 feet in 100. From the foot of the engine plane to the check-house, a distance of 1200 feet, the cars are taken over a level road by horses, and then in sets of three are lowered to the tipple on a self-acting inclined plane of 1500 feet length. On the return trip, when the set arrives at the drum, the latter is thrown out of gear, the rope is knocked off and replaced by a separate rope, worked by a separate drum, which pulls the train up an easy grade to the head of the incline, located inside the entry, about 200 feet from the pit mouth. The train-rider, who stands on the first car, drops the auxiliary rope at this point, and the set descends to the parting by gravity, checked by a brake on the drum. While the train, pulled by a separate rope, is passing over the drum, the engineer attaches the main rope to the rear car. The train itself, therefore, takes the rope on the other side of the drum and drags it into the pit.

The wooden supporting-rollers are six inches in diameter, placed twenty feet apart. The iron guiding-sheaves in the curves have a diameter of eight inches, and are similar to those previously described. The main rope and the rope of the self-acting incline have a diameter of ¾ inch, the auxiliary rope of ⅔ inch. They are steel ropes, with seven wires to the strand, and have been in service for several years without showing appreciable signs of wear. When descending the outer incline, a peculiar arrangement is used for keeping the rope low enough to drop into the guiding-sheaves. A 3x⅔-inch bar (*Fig. 23*), forged in the shape

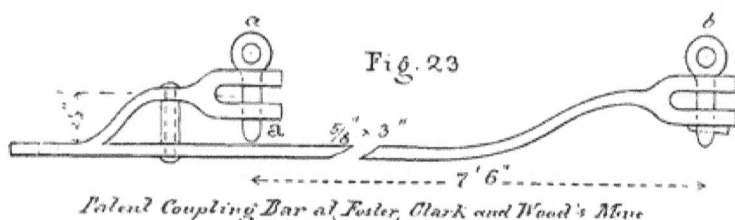

Fig. 23

Patent Coupling Bar at Foster, Clark and Wood's Mine

as shown, passes under the last car and is coupled with the pins (*a*) and (*b*) to the pulling-bars of the last and second last car. The rope is attached to the end of

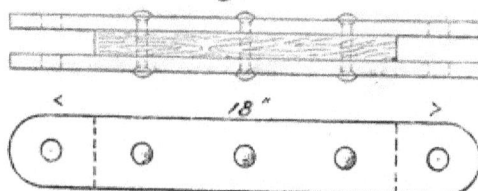

Fig. 24.

Rigid Car coupling used in the Coal Valley Mine.

this bar, which is two and a half inches lower than the regular pulling-bar. Arriving at the foot of the plane, it is dropped and thrown on the returning

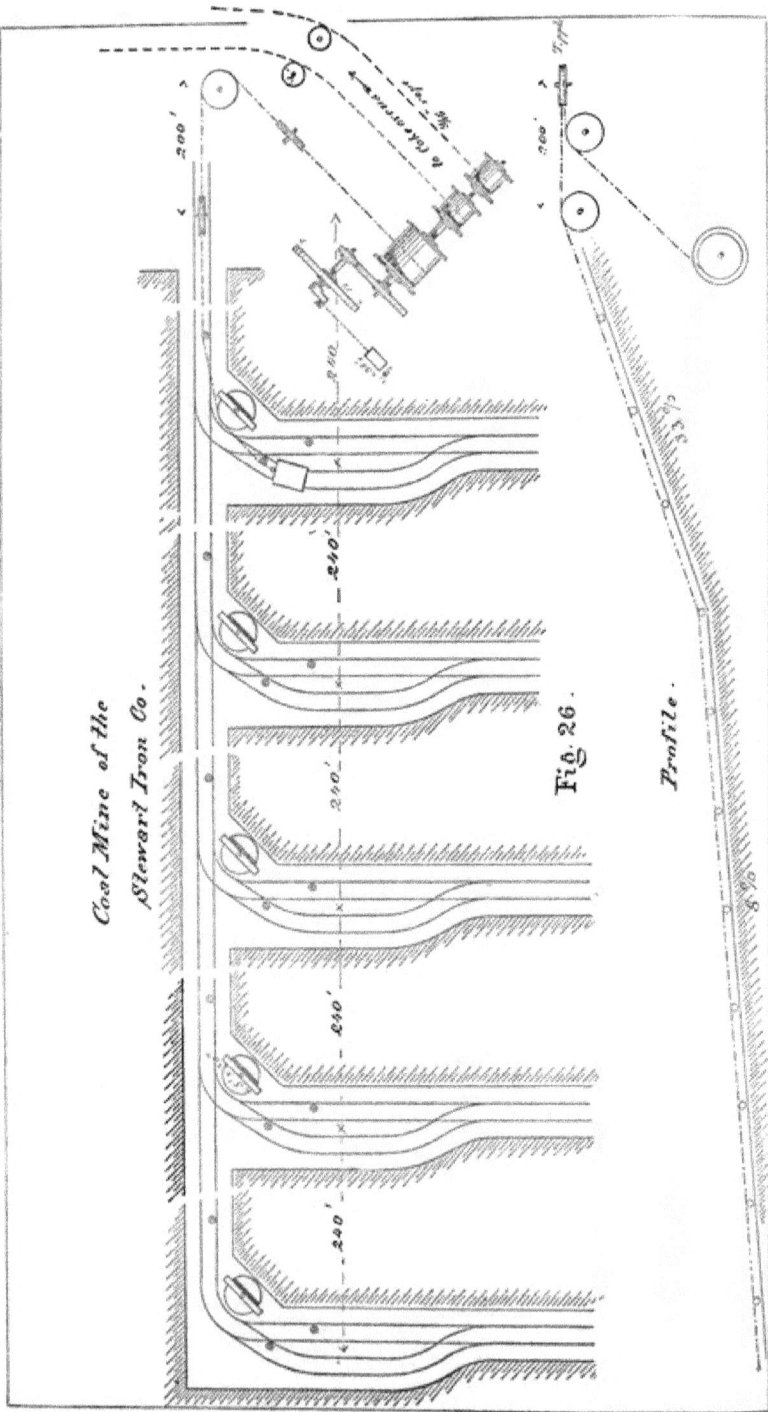

Coal Mine of the Stewart Iron Co.

Fig. 26.

Profile.

empty train. For connecting the cars a rigid coupling is used, consisting of two flat irons eighteen inches long, and a wooden filling piece riveted together (*Fig. 24*). The ends are provided with eyes for securing them to the pulling-bar of the cars with pins. A round trip is made in about thirty minutes, and the daily output in twenty trips amounts from 900 to 950 tons.

Of the many other engine planes distributed throughout the Pennsylvania coal region we will also mention those at the mine of the Mansfield Coal and Coke Co., where, by means of a ⅞-inch steel rope, twelve loaded pit cars are lowered on a long, steep and crooked plane from the pit mouth to the tipple.

At the Imperial mine an engine plane has been applied to replace a locomotive for shifting the railroad cars to and from the tipple (*Fig. 25*).

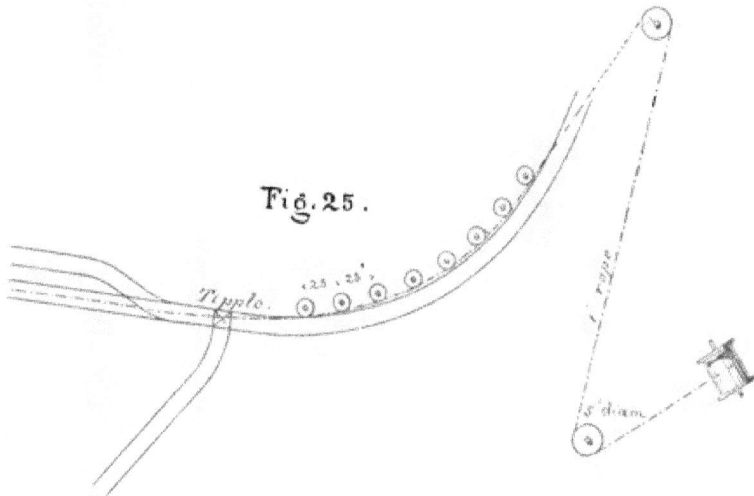

Fig. 25.

The cars are operated by means of a 1-inch steel rope and by the same engine which works an endless rope system for hauling the coal from the pit. The engine-house being located much higher than the railroad, and at considerable distance from it, it was necessary to turn two sharp angles with the rope by leading it around iron sheaves of five feet diameter. The rope has a length of 2200 feet, and a constant service of three years left it still in good condition. The railroad has a grade of one foot in thirty-two, and, as seen by the sketch, turns a sharp curve, around which the rope is guided by eight iron sheaves of 30-inch diameter, placed outside the track and twenty-five feet apart.

Another application of the engine plane principle is a wire rope arrangement for feeding coke ovens. A plant of this kind is in operation at the works of the Stewart Iron Co., which will be described in another chapter. At the same mine there is a slope for hauling coal from the pit to the tipple, operated in a different manner from those hitherto described. In place of a general parting at the end of the plane, for arranging the trains, there are five side entries,

240 feet apart, from which alternately sets of five cars are taken and returned (*Fig. 26, Plate 5*). At the junction of each entry with the main road, a wooden drum, of five feet diameter and twenty inches face, is placed outside the track and within six inches of the rail, for the guidance of the rope around the corner when being attached to the cars standing on the side track. By a mark on the rope, indicating the position of the set inside the pit, the engineer is able to stop the "empties" alternately at the different junctions. A man, passing from one junction to another, attends the switches, detaches and attaches the rope, and gives to the engineer a bell signal to hoist or to lower, as circumstances may require. The whole length of the rope is 1700 feet—1200 feet being used on the slope, 200 feet from the head of the slope to the tipple, and 300 feet from there to the drum below. It is made of steel, of one inch diameter, with nineteen wires to the strand. The grade of the slope from the top to the first side entry, a distance of 240 feet, is one foot in three, and from there to the end, eight feet in a hundred. The supporting rollers have a diameter of eight inches, and are placed twenty feet apart. A single car carries 2500 pounds of coal, and the quantity taken out per day is from 250 to 300 tons.

A safety arrangement, almost universally used on engine planes, where the load is raised, is illustrated in *Fig. 27, Plate 6*, in different shapes. It consists of an iron bar, two inches square and four and a half feet long, called the "growler," attached with a loose joint to the last car, so that it easily drags over the ground and rollers. If the rope should break, the pointed end of the growler digs itself in the ground and stops the downward motion or throws the cars off the track. Sometimes the growler is attached to the front car in lowering a loaded train down a steep incline. In this case a construction, as shown in *Fig. 27ᵃ*, may be employed. The iron bar is held high by a chain (s), which can be unhooked if the link (w) is pulled back by means of a light hemp rope running over the cars to the rear of the train, and operated by the train-rider.

The wire ropes for working an engine plane should in all cases have a safety factor of 5 to 6—that is, their breaking strength should be five or six times larger than the working strain. Where a breakage of the rope might cause great damage to property, or would endanger life, a factor of 6 to 7 is to be recommended.

III. THE TAIL ROPE SYSTEM.

Of all methods for conveying coal underground by wire rope, the Tail Rope System has justly found the most application. It can be applied under almost any condition. The road may be straight or curved, level or undulating, in one continuous line or with side branches—in all cases this system works with equal certainty and economy. In general principle a tail rope plane is the same as an engine plane worked in both directions with two ropes. One rope, called the "main rope," serves for drawing the set of full cars outward ; the other, called the "tail rope," is necessary to take back the empty set, which on a level or undulating road cannot return by gravity. The two drums may be located at

Pl. 6

Fig. 17

Hand break

rigid coupling

Fig. 17ᵇ

loose coupling

Fig. 17ᵃ

Forms of Pit Cars.

Fig. 27ᵃ

Fig. 27ᵇ

Fig. 27ᶜ

Fig. 27ᵈ

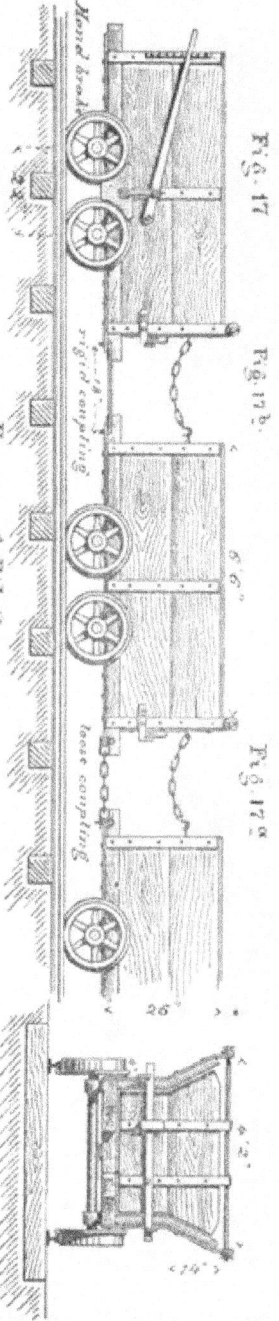

Various forms of "Growlers"

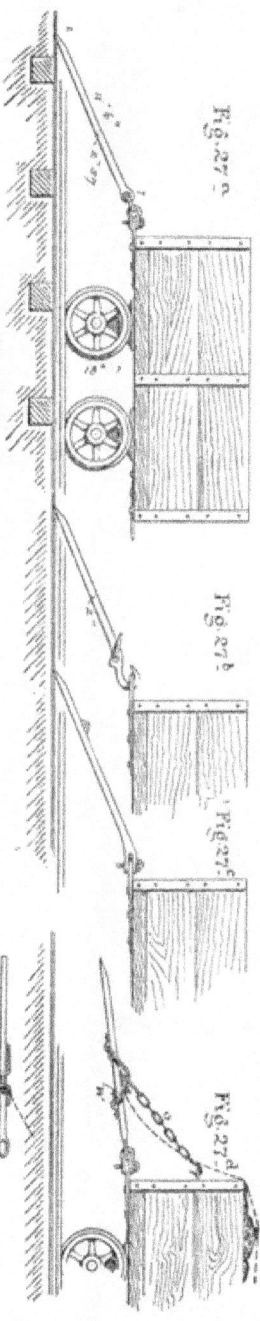

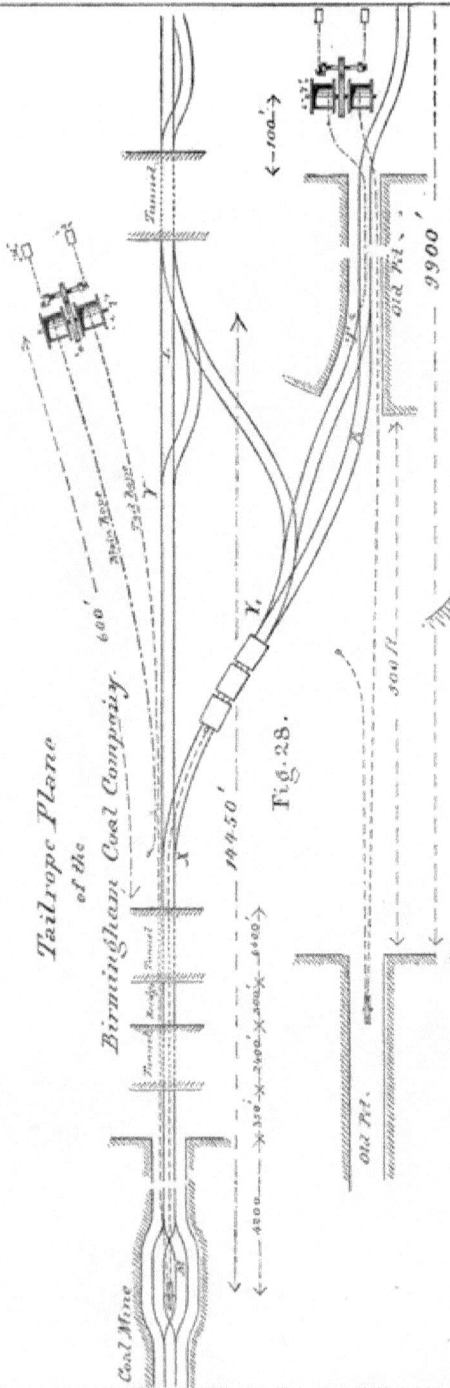

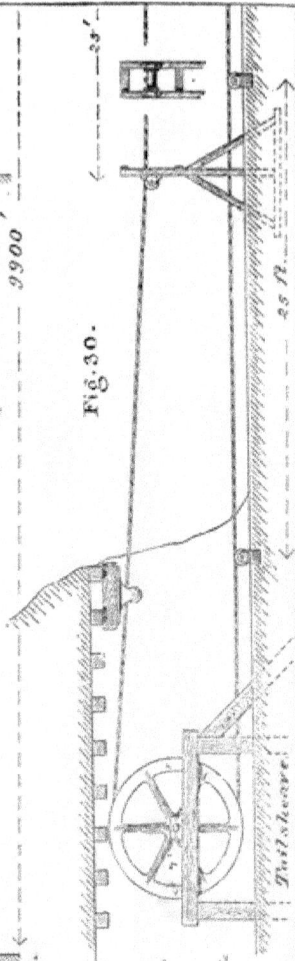

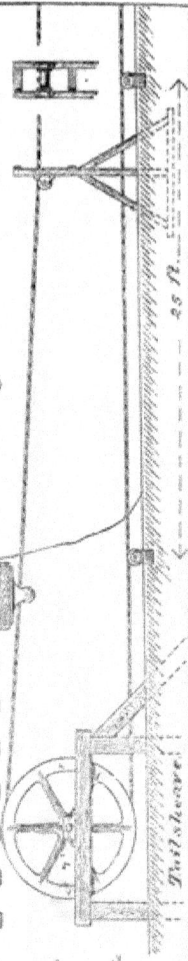

Tailrope Plane
of the
Birmingham Coal Company.

Fig. 28.

Fig. 29.

Fig. 30.

Tailsheave.

the opposite **ends of the** road, and driven by separate engines, but more **frequently** they are **on the** same shaft at one end of **the plane.** In the first case each rope **would** require the length of the plane, but in the second case the tail rope must **be** twice as long, being led from the drum around a sheave at the other end **of** the plane and back again to its starting point. When the main rope draws a set of full cars out, the tail rope drum runs loose on the shaft, and the rope, being attached to the rear car, unwinds itself steadily. Going in, the reverse takes place. Each drum is provided with a brake to check the speed of the train on **a** down grade **and** prevent its **overrunning** the forward rope. As **a** rule, the tail rope is **strained** less than the main rope, **but** in cases of heavy grades dipping outward it **is** possible that the strain in **the** former may become as large, or even larger, than in the latter, and in the selection of the sizes reference should be had to this **circumstance.**

A description of **a few** tail rope planes will more fully explain **this system.** *Fig. 28, Plate 7,* illustrates the general plan of the extensive establishment **of** the Birmingham Coal Co. It **consists of two** planes, joining each other, one 14,450 feet, **the** other 9900 feet long, each worked by a main and tail rope. In connection **with** them there are two gravity planes of about 1800 feet length, so **that it requires** an aggregate length of over fourteen miles of **wire rope for** transporting **the** coal from the parting **within** the **pit** to the **tipple.** The road is straight, **with** undulating grades varying **from** ½ to 2 feet in 100, **and** takes its course from the mine through **open** country, over bridges and through various tunnels, the trains moving at a speed of eight miles per hour. A **set,** consisting of **from** 60 to 75 loaded cars, starts **from track** (M) of the parting, the main rope being attached to the front, the tail rope **to the rear** of the train. When arriving at point (X) **the main** rope is knocked **off** by a man stationed there, and the train runs on a light down grade alternately to track (L) **or** (R), the tail rope being knocked off when the last car reaches points (Y) or (Y_1).

The former trains **are** taken by a locomotive **to** one shipping station, the latter to another station **by a** second **main and** tail rope attached to the ends of **the** train at the points (T) and (T_1). On the return trip **of** the empty **set these** ropes are detached at the same points and the tail rope of the other **plane is** hitched to the forward part of the train at (Y), or at (Y_1) if the set **arrives** by locomotive, while the main rope, without stoppage, is coupled **to the rear car** at (X) when passing this point. All coupling and uncoupling of **ropes,** switching and giving signals at the junction of **the** two planes, **is done by one** man, while another one attends to the same duties at the parting. The signals are given by telephone, and no train-rider accompanies the trips. **A** full set has a weight of 120 to 150 tons, a single car carrying 1¼ tons of coal, and weighing, empty, ¾ of **a** ton. Its dimensions are 6 feet 4 inches long, by 28 inches wide **at the** bottom and 42 inches **at the** top, and 30 inches deep. The wheels have a **diameter of 18** inches, running loose on 2½-inch axles, which are placed two feet **apart.** The car-coupling consists of **a** short chain, with a clevis at each end, **secured** to the pulling-bar, and the usual two safety-chains (*Fig. 17*, *Plate 6*). The main **rope** has a diameter of ⅞ inches, the tail **rope of** ¾ inches; both are of steel, with seven wires to the strand, and were still in good condition after several years' service. Their attachment consists of a goose-neck socket and clevis, with a single intervening link (*Fig. 29, Plate 7*). The main rope is supported in the middle of the track by wooden rollers of 8-inch diameter and 24 inches long, placed at distances of 25 feet; the "empty" tail rope runs alongside the track over rollers of 6-inch diameter and 11 inches long, supported from the roof or side wall of the entries, or outside of them, resting on specially erected frames, as illustrated in *Fig. 30*.

This sketch represents, also, the upright "return-wheel" for the tail rope at the end of the lower plane. It has a diameter of 7 feet, and consists of a grooved iron rim, lined with wood, but otherwise not differing in construction from an ordinary heavy iron hoisting-wheel. The rope drums have a diameter of 7 feet, and can be put in or out of gear by a "V clutch." Each system is driven by an 80-horse power double-acting engine, with two cylinders of 14-inch diameter by 36-inch stroke, making 65 to 70 revolutions a minute, a 5-foot pinion and 12-foot spur-wheel, with 9-inch face. A round trip occupies 45 minutes, and the daily output averages 1200 tons.

A tail rope system, with drums driven by separate engines and located on the opposite ends of the plane, is represented at the works of the Mansfield Coal and Coke Co. (*Fig. 31, Plate 8*). The road is 4900 feet long, with no uniform grade and with two sharp curves forming an S inside the pit. It is worked by a ⅞-inch steel rope, both for main and tail rope. The drum (*A*) for the former is located near the pit mouth, while the tail rope crosses the track under the rails at (*L*), not far from the return-wheel, and is led through a separate entry of 1300 feet length to drum (*B*) at the mouth of this entry. Both drums have a diameter of 5 feet; each is worked by a 30-horse power single-acting engine, with a 12x30-inch cylinder, 8-foot fly-wheel, 18-inch pinion, and 4-foot 10-inch spur-wheel on 6-inch shafts. An extra drum (*C*), placed on the same shaft with the main drum, serves for working the engine plane, on which the full cars coming from the pit are lowered to the tipple, and which has been mentioned already in the preceding chapter. The following is the mode of operation : A set of 24 full cars starts from track (*M*) of the parting, with main and tail rope attached to the first and last car, and is pulled out with a velocity of 10 miles per hour. As soon as the first car comes outside the pit the main rope is knocked off and the train allowed to run by its own momentum as far as (*P*), at the head of the engine plane. When the last car has emerged from the pit the tail rope is detached at point (*Q*) and replaced by the ⅞-inch rope of the engine plane. This latter rope passes around a 7-foot sheave in order to bring it in the direction of the road. A "lock" at point (*P*), which prevented the premature descent of the cars, is now opened, and the train is let down, checked with a brake on drum (*C*). On the return trip the different operations are repeated in reversed order. The last car of each train, ingoing as well as outgoing, is a special car, called the "dilly" (*Fig. 32*). It consists of a small truck, loaded with metal, and the coupling-bar low enough to hold the rope down and guide it into the sheaves when going around the curves. For the easier curves, ordinary iron

Tailrope Plane of Grey and Bell's Mine No 1.

Fig. 31

Fig. 32

The "Dilly" of the Mansfield Mine.

Fig. 33.

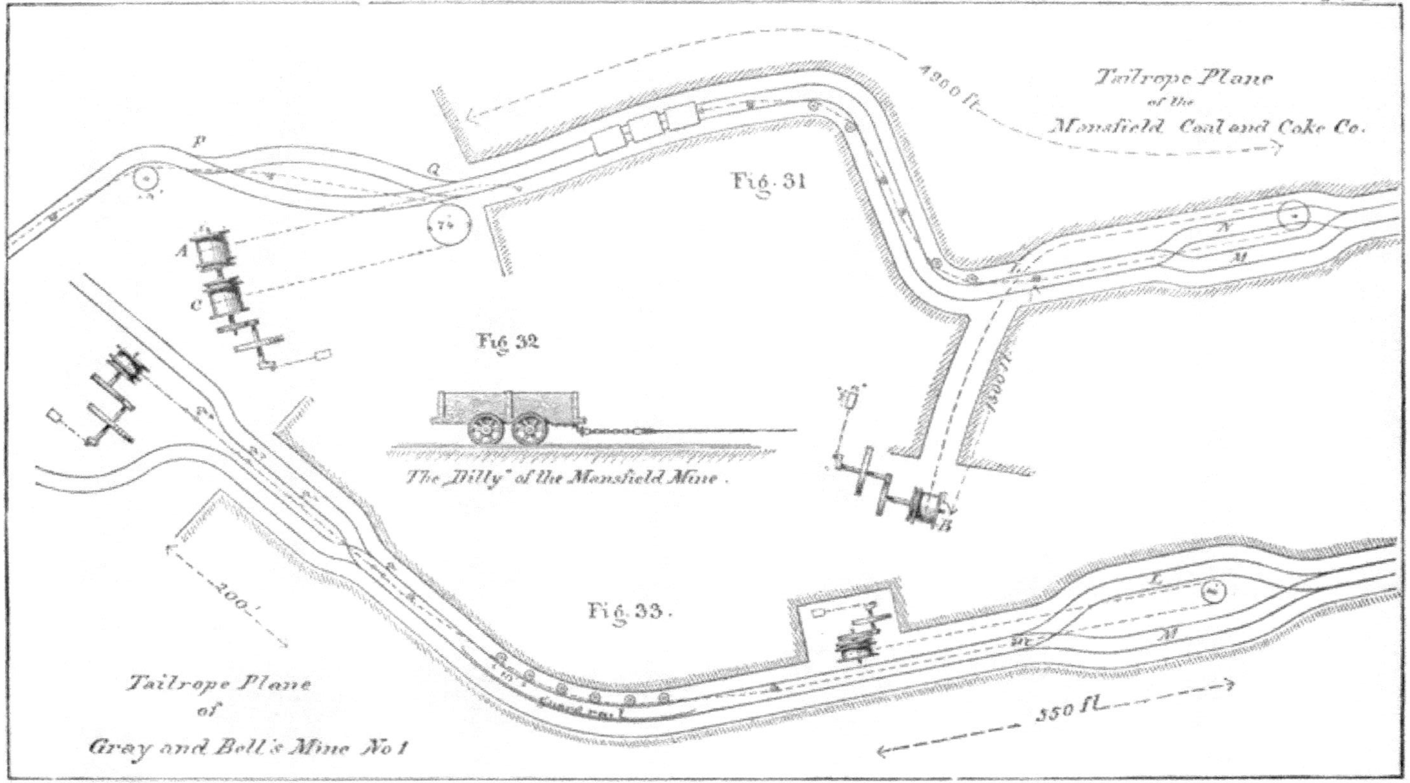

Tailrope Plane of the Mansfield Coal and Coke Co.

100 ft.

490 ft.

750 ft.

350 ft.

p. 8

sheaves of 10-inch diameter are used, placed inside the track, but at the two sharp turning points a number of old pit wheels of 18-inch diameter are placed outside the track on the concave side of the curve. As this produces a sideward pull, it is necessary to provide these points with guard-rails to prevent the cars running off the track. At either end of the road the dilly is left on the parting and pushed by hand to the rear of the train after the latter has reached the main track. A signal is given to the engineers for one to hoist and the other to watch the loose drum and keep the dragged rope taught, but no train-rider accompanies the trip.

The coal works of Gray & Bell employ, at their different mines, a combination of nearly all wire rope systems for conveying coal. *Fig. 33, Plate 8*, represents a general plan of one mine which is worked by a main and tail rope upon the principle just described. The drum and engine for the main rope are near the pit mouth, while the tail rope drum is located inside the pit, about 550 feet from the return-wheel, at the foot of a ventilating shaft. The road has a length of 6600 feet, with a total fall of 120 feet, the average grade, dipping inward, being 15 inches in 100 feet, but at some places 4 or 5 feet in 100. It consists of two straight lines forming an angle of 106 degrees, around which the rope is guided in a sharp curve by means of six iron, wood-lined sheaves of 18-inch diameter, placed ten feet apart outside the track, on the concave side of the curve. The wood filling in these sheaves has to be replaced about every three months. Guard-rails are placed at this point for the guidance of the cars. Unlike the other tail rope systems, where the rope ends are attached to the ends of the train, the two ropes of this plane are connected and therefore practically made endless. The connection consists (*Fig. 34, Plate 9*) of a chain four feet long, attached to the ends of the ropes by means of goose-neck sockets. It contains, in the middle, a solid cylindrical link, with a loose ring on it, to which the cars are hitched with a 10-foot coupling-chain. The loose ring admits a free turning of the rope, replacing a swivel in the chain, and answering the same purpose fully as well, or even more effectively. There are four such couplings, placed at a, a_2, a_3 and a_4, at such distances that when the first car of a train is at a, the rear car reaches either to a_2, a_3 or a_4, according to the number of cars in the train, which always is composed either of 25, 36 or 40 cars. The latter number, being the maximum for one train, represents a weight of 80 tons, which is taken by a $\frac{5}{8}$-inch steel rope over a distance of 6600 feet in fifteen minutes, and lifted to a height of 120 feet. Calculating the strain on the rope for this load, when ascending a grade of 5 per cent., we find that it will amount to 6 tons, or more than one-half of its breaking strength. In consequence of this excessive strain with which the ropes are taxed, one year's service is considered a sufficient duty. The main rope is replaced every year by a new one, and for the following year it is used as tail rope, and after two years' service is entirely discarded. The following is the method employed for working this plane: No coal is mined between the pit mouth and return-wheel; it is collected by mule-drivers from the interior entries of the mine, and arranged as a set of 25 to 40 cars on track (*L*) of the parting. The first car is hitched to the main rope, the last car left free, but provided with a safety-bar as illustrated in *Fig. 27ᵃ, Plate 6.* On a bell signal given to the engineers the main rope winds up and the tail rope unwinds itself, the train moving out accompanied by a train-rider on the front car. When within 15 or 20 feet of the pit mouth the engine is stopped, the train-rider pulls the coupling-pin of the chain, letting the train run by its own momentum to the tipple. The short distance from the tipple to the pit mouth makes it necessary to take the empty cars, with a mule, to the end of the switch inside the pit, about 200 feet from the entrance. The first car of the empty set is now hitched to (a), these rope-couplings being arranged so that they never go around the drum nor return-wheel, and always stop at the same points between (a_4) and (m). The rear car is also hitched to one of the other couplings, to prevent the cars from overrunning each other on the steep down grades. When arriving at (m) the train-rider pulls the coupling-pin and runs the train on the right hand track (M) of the parting, and the whole operation is repeated with the next set. The advantage of the endless rope consists, first and principally, in the fact that the rope always runs in the sheaves and rollers, and does not require a "dilly," "dilly-bar" or other arrangement to guide it in the sheaves ; secondly, it saves for the outgoing trains the work of coupling the last car. The combined length of main and tail rope is 14,000 feet, and both branches are supported every 21 feet by 6-inch wooden rollers, which last from five to six months. The return-wheel has a diameter of 4 feet, and its rim has a wood filling which has not been replaced in ten years. The main rope drum has a diameter of 5 feet, with a 20-inch face; it is driven by a single-acting engine, with a 16-inch cylinder, having a 24-inch stroke, a 5-ton fly-wheel and a gearing of 1 to $2\frac{1}{2}$. It has about 50 per cent. more power than necessary to pull a train of 40 cars from the parting to the pit mouth at a rate of five miles per hour. The average daily output is 1000 tons of coal.

Another wire rope plant, owned by the same company, for transporting coal over a distance of 13,390 feet, is illustrated in *Fig. 35, Plate 9.* It consists of two tail rope systems, an engine plane and a self-acting plane, all worked by $\frac{3}{4}$-inch steel ropes. The coal, collected from the mine back of the parting (L), is taken in trains of 28 cars by a tail rope system to the pit mouth, from here by a single rope and engine plane through an old pit to the top of the hill, then in single cars down a self-acting incline of 1320 feet length, and finally in sets of 46 cars by a second tail rope system through a tunnel to the shipping station.

A little variation will be noticed in the position of the drums of the upper tail rope plane. Instead of being on the same shaft they are placed behind each other, which necessitates a second return-wheel for the tail rope close to the drum. The diameter of the drum is 5 feet. An extra drum of 4-foot diameter is placed on the same shaft of the main drum for working a $\frac{1}{2}$-inch rope, which is attached to the rear car of the train when it ascends the engine plane to station (C) At point (F), about in the middle of the tunnel, the train-rider drops this rope, which afterwards serves to haul the empty cars from the foot of the engine plane back to station (B). The gravity inclined plane, as operated in this plant, has already been described in the first chapter, and is represented in *Fig. 2C,*

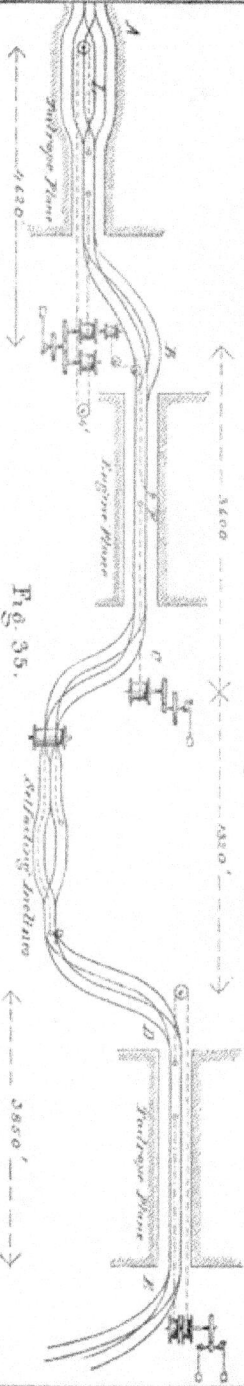

The Coalworks of Gray and Bell. N.º 2

Tailrope and Engine Planes

Fig. 35.

Fig. 34.

Profile

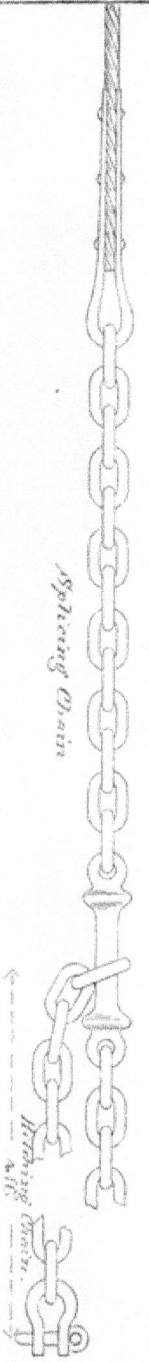

Splicing Chain

Pl. 5

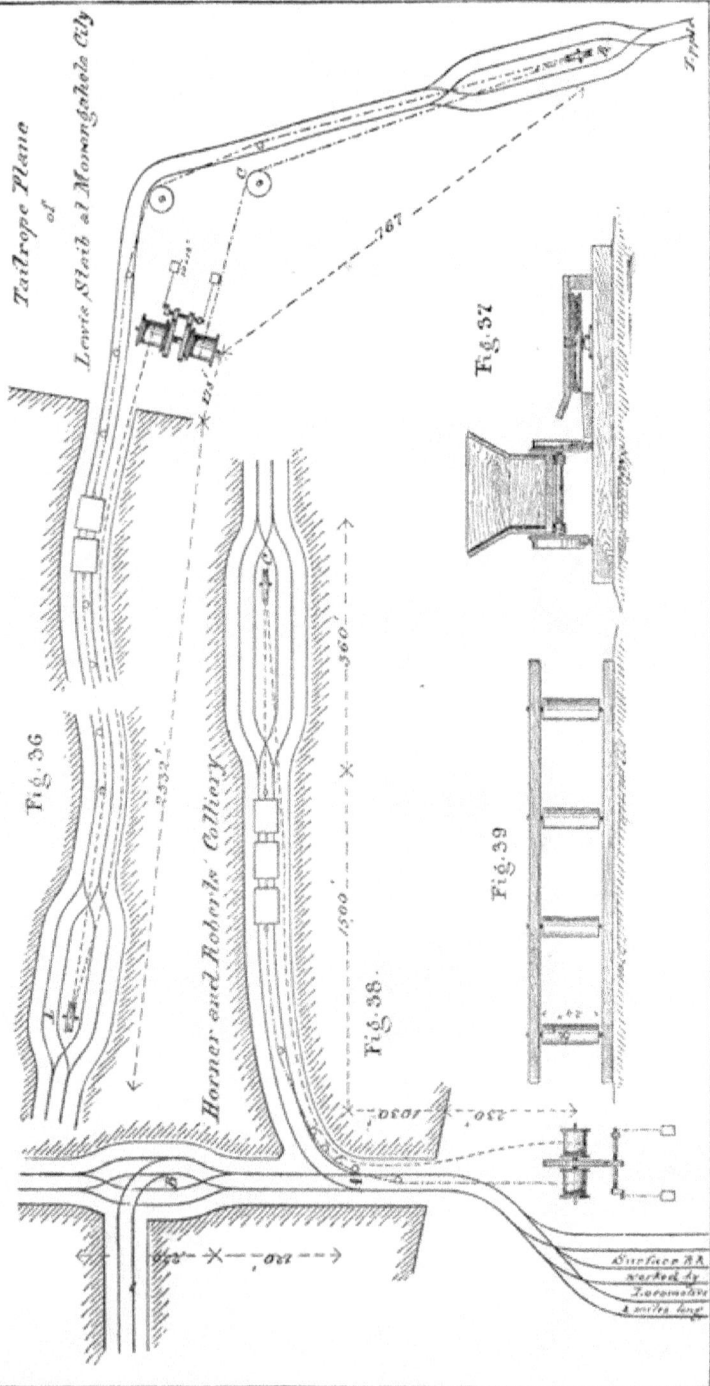

Pl. 10

Tailrope Plane
of
Lewis Slit at Monongahela City

Fig. 36

Horner and Roberts Colliery

Fig. 37

Fig. 38.

Fig. 39

Surface RR
worked by
Locomotive
4 miles long

Plate 1. At the mouth of the last tunnel there is not sufficient room between tunnel entrance and drum to haul the whole train out; therefore, as soon as the first car emerges from the tunnel it is detached from the main rope, the tail rope drum is put in gear, the main rope drum out of gear, and the coupling of the main rope hauled back to the last car. To this it is hitched again, so that after reversing the gear of the two drums the train finally is pushed out. The daily output of this mine is 21,000 bushels, or 798 tons, and the owners estimate that 150 mu'es and the corresponding number of drivers would not be sufficient to perform the same work as is done by the rope systems in the two mines, while the wear and tear of the ropes, with the maintenance of rollers, sheaves, &c., does not exceed one cent per ton.

Another variation of the tail rope system, illustrated in *Fig. 36, Plate 10,* represents the arrangement at Lewis Staib's mine, on the Monongahela river. Though the drums are not located at the end of the plane, it is managed by means of two return-wheels, one at the parting (*L*), the other at the tipple (*B*), to convey the coal between these two points in one uninterrupted pull. The main rope is supported, as usual, in the middle of the track, on 6-inch rollers, placed 18 feet apart, while the tail rope runs overhead on rollers supported from the roof of the pit. From the engine-house to the tipple the road runs over a trestle. The empty main rope is led under the rails in order to clear the crossings, and from point (*C*) it is stretched to the drum without any support. *Fig. 37* shows the sheaves at the sharp turning-points of the road. The engine is double-acting, with 10x12-inch cylinders, 4-foot drums, and a gearing of 1 to 5½. The road dips inward, with various grades, averaging 2 feet in 100, and has many curves around which the rope is guided by 12-inch wooden sheaves. A round trip from the parting to the tipple and back again occupies twelve minutes, and the daily output of coal amounts from 750 to 900 tons. The main rope has a diameter of ⅝ inch, the tail rope of ⅜ inch; both are of steel, with seven wires to the strand. The total cost of the plant, including engine and engine-house, is estimated at $8000. It furnishes an excellent example of the adaptability of the main and tail rope principle under very inconvenient circumstances.

A wire rope plane, almost identical with this in general arrangement, is also in operation at the mine of James Jones. The drums are located 850 feet from the return-wheel at the tipple and one-half mile from that at the parting. About 400 tons of coal are mined per day, and conveyed over the distance of 3500 feet by a ⅝-inch main and ½-inch tail rope, which replace the service of twelve mules and six drivers.

Fig. 38, Plate 10, representing Horner & Roberts' colliery, shows an example where two branches are worked by a main and tail rope system. The road dips from the pit mouth to the parting (*B*) at a rate of 5 feet in 100, but from the junction of the two branches at (*A*) to the parting of the side entry it is almost level. If an empty set is intended for the side branch, it is taken in by the usual way with tail rope in front and main rope in the rear of the train. Returning from this branch the tail rope is hitched to the second last car, the last one being provided with a growler (*Fig. 27ᵇ, Plate 6*). If, however, a set is intended for the parting (*B*), the train-rider, who accompanies every trip, gives to the engineer a signal to stop at the junction of the branches. He detaches the tail line, throwing it out of the way to one side of the track, and lets the train descend by gravity to the parting (*B*). When returning with the full trip another stop is made at the junction for the purpose of attaching the tail rope again to the rear car. As the growler would interfere with this coupling if secured to the last car, it is hitched to the second last car on all trains leaving station (*B*), and the last car is coupled only with the safety-chains. The road to (*B*), therefore, is nothing but a simple engine plane, worked by the main rope of the tail rope plane. A regular full trip consists of 40 cars, carrying 1600 bushels, equal to 60 tons of coal, and representing a total weight of 86 tons. About fifteen trips are made per day, giving an output of 900 tons. Both main and tail ropes are of steel, with seven wires to the strand, the first of $\frac{7}{8}$-inch diameter, the latter of $\frac{3}{4}$-inch. The supporting-rollers are placed 18 feet apart, those for the tail rope being supported from the roof of the entry. The rope attachment consists of an ordinary socket and a heavy chain eight feet long, without swivel. The weight of the chain helps to keep the rope down and facilitates its dropping in the sheaves at the curves. At the junction of the two branches, four wooden rollers, of 8-inch diameter by 24 inches long, are placed in upright positions outside the track, to guide the ropes around the corner (*Fig. 39*). Both ropes have been in use three and one-half years without showing noticeable signs of abrasion. They receive, occasionally, a coating of a mixture of tar, oil and finely-ground burnt lime, which helps to preserve them. The maintenance expense of rollers and sheaves does not exceed ten dollars a year.

Another notable tail rope system is at the Smithton mine, on the Youghiogheny river. The plane has a length of 1$\frac{1}{4}$ miles, with an undulating grade (*Fig. 40, Plate 11*) and many curves. It is worked by a $\frac{3}{4}$-inch steel rope, supported every twelve feet by 6-inch wooden rollers, and guided around the curves by 12-inch wooden sheaves placed between the rails. These sheaves (*Fig. 41*) are composed of three 1$\frac{1}{4}$-inch boards, bolted together between two iron flanges, and secured to the ties by a bolt 1$\frac{1}{2}$ inches in diameter, which at the same time serves as an axle. The tail rope, after passing around the return-wheel, is led through an air-course back to the drum, and therefore is entirely out of the way on the main road. In front of the pit mouth the principal track is joined by a side track leading to the entry of another mine, in which all hauling is done by mules. They bring the coal to this junction (*Fig. 42*), from whence it is conveyed to the tipple by the wire rope system in the following manner: An empty set, on its return trip to the lower pit, is temporarily switched on track (*L*); the two ropes are detached from it and hitched to a full set standing on track (*N*), taking it to the tipple and returning with another empty train, leaving the latter on track (*M*). Then the ropes are changed back again to the first train, standing on (*L*), which is hauled without further interruption to the parting inside the pit. All cars used in this mine are provided with hand-brakes, after the pattern shown in *Fig. 17, Plate 6*, on account of the steep grades in the entries and rooms. A regular load consists of 40 full cars, and occasionally of 65. In the centre of

Pl. 11

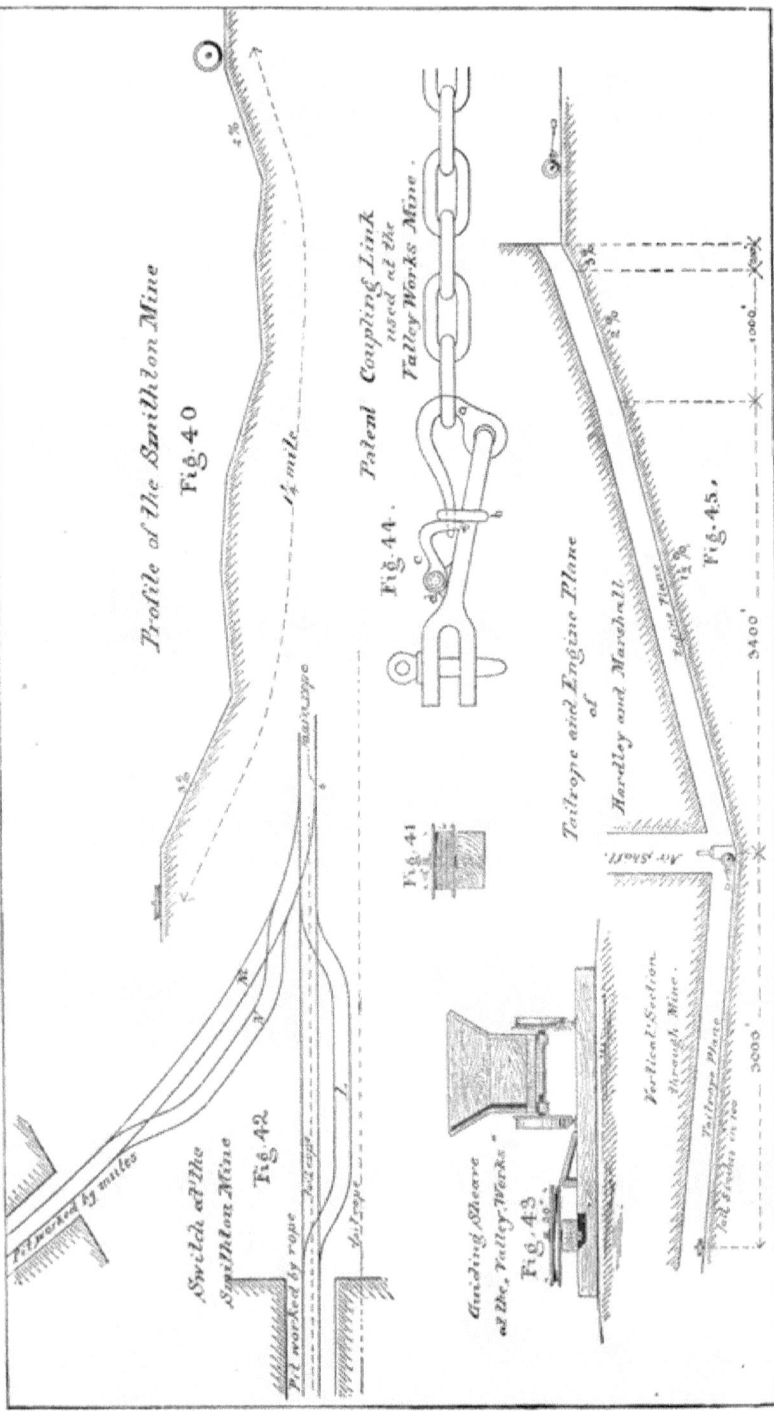

Profile of the Swithlon Mine
Fig. 40

Switch at the Swithlon Mine
Fig. 42

Pit worked by mules

Patent Coupling Link
used at the
Talley Works Mine.

Fig. 44.

Fig. 41.

Guiding Sheave
at the Talley Works
Fig. 43.

Tailrope and Engine Plane
of
Hartley and Marshall.

Fig. 45.

Vertical Section
through Mine.

Air Shaft.

Tailrope Plane

the inner parting there is a row of posts to support the roof of the pit, and in consequence of these it is necessary to change the trips alternately from the left-hand to the right-hand track of the parting. The coal vein has an average thickness of 9 feet, and the daily output of coal is 1140 tons.

The Valley works have a tail rope plane similar to the preceding one. It is 4400 feet long, and conveys per day 950 tons of coal by means of a $\frac{3}{4}$-inch main and a $\frac{1}{2}$-inch tail rope. The grade is undulating, varying from 1 to 4 feet in 100 against the load. *Fig. 43, Plate 11,* shows the arrangement for guiding the rope into the sheaves at the curves. The sheave is placed outside the track, a little higher than the coupling-bar; an inclined piece of wood, $2\frac{1}{2}$ inches square, running from the rail to the flange of the sheave, prevents the rope from dropping under the same and guides it into the groove. An ingenious coupling (*Fig. 44*) is used for the attachment of the ropes to the cars, enabling the train-rider to unhitch the rope at any time or point without stopping or slacking the engine. It consists of a movable hook (a), held in position by a ring (b), which is prevented from slipping out by a prong (c). The latter, turning around the pivot (d), can easily be lifted up, allowing the ring to slide back. This frees the hook, which by the pull of the chain turns over and unhooks itself.

At the coal works of Joseph Walton and Thos. Fawcett, on the Monongahela river, are three other examples of large tail rope planes. Each plane has a length of $1\frac{1}{2}$ miles, and is worked by a $\frac{7}{8}$ to $1\frac{1}{8}$-inch main, and $\frac{1}{2}$ to $\frac{5}{8}$-inch tail rope. In the two Walton mines about 1500 tons, and in the Fawcett mine 500 tons, are daily mined and transported the above distance by a tail rope system. The general arrangement and method of operation is the same as described before in such examples as where both rope drums were placed on the same shaft at one end of the plane.

A tail rope system in the mine of Hardley & Marshall, in conjunction with the engine plane described in the former chapter, is especially notable for the favorable conditions of the road, on which any other motive power for transporting coal could be employed with advantage. Nevertheless, the preference was given to a wire rope system, which even in this case proved to be the most convenient and cheapest possible method. The road is 3000 feet long, and has a uniform grade of 5 inches in 100 feet in favor of the load. The engine and boilers are located in the pit, at the foot of a ventilating shaft, at the junction of the tail rope and engine plane (*Fig. 45, Plate 11*). Both main and tail ropes are of steel, of $\frac{7}{8}$-inch diameter; the latter is led from the return-wheel to the drum through an air-course, which runs, at a distance of 24 feet, parallel with the main entry. The supporting-rollers for the main rope, placed between the rails, are 27 feet apart; those for the tail rope are supported from the roof of the air-course, and placed at distances of 90 feet from each other. The daily output of coal is 950 tons, and the cost of transporting the same over the whole distance of 7600 feet, by wire rope, amounts to $1\frac{1}{4}$ cents per ton, which includes all the necessary labor and maintenance of the road and machinery. Less than one-fourth cent of this sum must be calculated for wear and tear of ropes and rollers.

Two examples of a combination of a tail rope with an engine plane, worked by one engine, are represented at the Lovedale works and at the mine of Gumbert & Huey, on the Monongahela river. The two plants are almost identical, with the exception that the engine plane of the first-named works has a length of 4374 feet ; the latter of 6000 feet (*Fig. 46, Plate 12*). From the pit mouth (*B*) to the check-house (*A*) there is a fall of $1\frac{1}{2}$ per cent., but in opposite direction, to the parting inside the mine, the road descends at a rate of 7 feet in 100 feet. Both rope drums are on the same shaft, driven by a double-acting engine with two 14x24-inch cylinders, the boiler carrying 100 pounds of steam. The tail rope runs from the drum around a return-wheel placed under the rails inside the pit about 200 feet from the mouth, and its end is attached to the front car of an empty set standing at (*A*), pulling the same up the incline to the entrance of the pit. While the train runs past the drum the main rope is hitched to the rear car, and when arriving at point (*C*), near the return-wheel, the train-rider drops the tail rope and the set descends with the main rope in tow to the inner parting by force of its gravity. On the return trip of the full set a stop is made near the return-wheel to enable the train-rider to hook the small rope in the ring of the growler (*Fig. 27°, Plate 6*), and to walk from the rear to the front of the train, taking his position on the first car. When outside the pit, at point (*F*), the engine rope is dropped and the train runs alone to its original starting point at (*A*), where the tail rope is changed to another empty train to repeat the same operation. A full set consists of 36 cars, and a round trip occupies 25 to 30 minutes, making the daily output of coal in each mine 500 to 600 tons.

We have seen that in all wire rope planes, so far described, the coal was collected at one end of the main road and conveyed to the other end without interruption, and in most cases without a stop. In the branches and side entries the coal was hauled by mules and brought by them to the general parting. There would be no difficulty in working one or more branches also by wire ropes, and it is in common practice in the collieries of England, though it has not yet been attempted in this country. This is the more to be wondered at, as in many cases the side entries are very long, and a wire rope arrangement for transporting coal in them would prove as advantageous as it is now on the main roads. No extra machinery is necessary, and with little additional cost it is possible to extend an already existing tail rope system to a number of branches. The principal features to be considered for working several planes with one main and tail rope are the following : Each branch road is provided with a separate rope, resting on the usual supporting-rollers and passing around a return-wheel at the end of the plane, both ends of the rope reaching to the junction of the branch with the main road. The rope on the main road consists of as many pieces as there are branches, the connections being made by sockets and shackles, and arranged at such distances that when the train is at the outer end of the plane, these couplings are in every instance opposite the junctions of the branches with the main road. If a set is intended to be drawn into one of the side entries, it is only necessary to disconnect the main rope from its upper part and connect it with

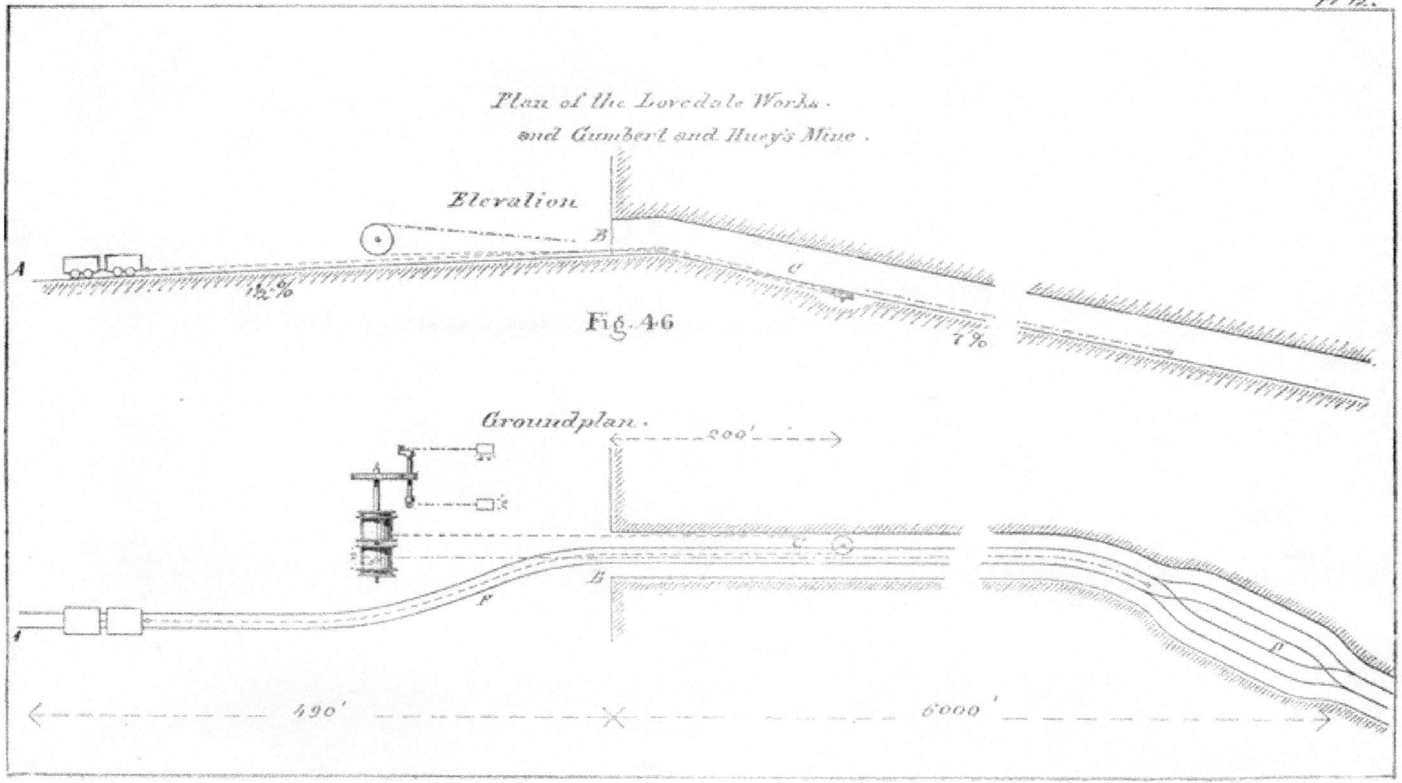

Plan of the Lovedale Works.
and Cumberland and Harp's Mine.

Elevation

Fig. 46

Groundplan.

the branch rope, so that the latter forms a continuous line **from its return-wheel to the drum**, while the remaining part of the main rope, as well as all other branch ropes, lie idly on the ground. There are three different methods for attaching the branch rope ends with the main rope, illustrated in *Fig. 47–49.*

Fig. 47. Fig. 48 Fig. 49.

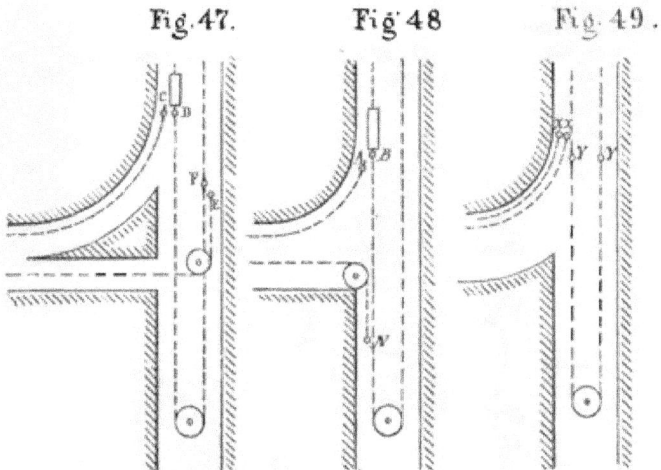

Method of attaching the Branchrope
with the Mainrope.

In the first two sketches the ropes are changed when the set of cars is near the branch end, and on the latter sketch when the set is at the outer end of the plane. In *Fig. 47* a wheel is fixed near the roof or under the rails, around which one end of the branch rope passes. When the incoming set has to go into this branch, the rope end (C) replaces (D) on the fore end of the set, and the end (E) replaces (F) on the tail rope. In *Fig. 48* the tail rope always remains entire;

Fig. 50

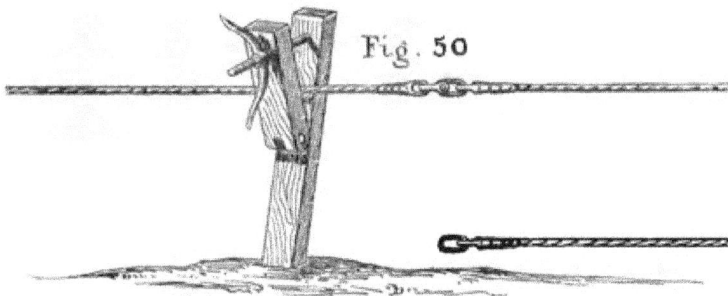

Clamp for holding the Mainrope
while attaching the Branchrope.

the end (A) replaces (B), and the end (B) of the tail rope is brought a little further by the engine, and is then attached to (N). A different course is pursued by the method shown in *Fig. 49*. Whilst the ropes are changed at the end of the plane from the full to the empty train, a boy at the branch end simply replaces the ends (XX) by (YY), and the train runs into the branch without stopping. This plan is more expeditious than either of the others, since no time is lost by stopping at the branch ends, and the ropes are changed while there is no strain on them. In the first two methods, if the road is not level there may be a heavy strain in the ropes, according to the dip of the road, and it is therefore a common practice to hold the ends of the main tail rope by a wooden clamp (*Fig. 50*) fixed at the junction to prevent the rope from rebounding when it is released from the strain. For uncoupling the ropes, when under a strain, it is necessary to use so-called "knock-off links," of which *Figs. 51* and *52* represent two kinds used in the English collieries. The cotter (C) in either of these is

Fig. 51

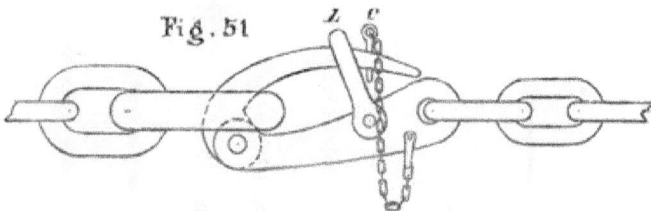

"Knock off link" used at North Hetton colliery.

Fig. 52.

"Knock off link" used at Harraton Colliery England.

removed and the link (L) is easily pushed off with the foot. In some mines the main rope at the fore end of an outgoing set is provided with a self-acting knock-off link (*Fig. 53*). When the car arrives at the place where the rope should be taken off, a piece of iron fixed to the roof of the pit or to a frame outside of it comes in contact with the arm (A) of the knock-off apparatus and releases the main rope, which falls to one side.

It will be noticed that it requires, altogether, eight disconnections and connections to run a train from the main road into a branch or reverse, and twelve couplings and uncouplings for running a set from one branch into another branch, namely: four at the **outer** end in changing the ropes from the full to **the empty set,** four at the junction of the first branch to release the branch rope ends and restore the continuity of the main rope, and four at the second junction

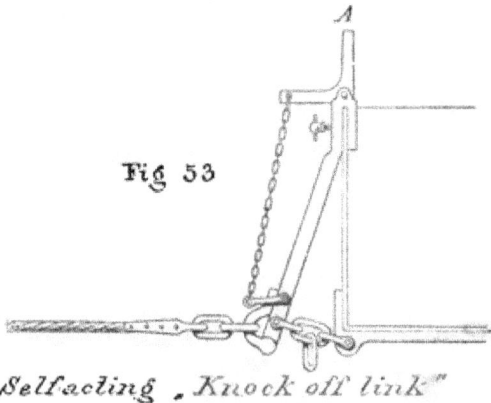

A

Fig 53

Selfacting „Knock off link"

to intercept the main rope at this place and connect it with the branch rope. In the methods of *Figs. 47, 48* the connections can be made by the train-rider, but in that of *Fig. 49* a boy is needed at the junction to make the changes and a train-rider could be dispensed with, though the latter is generally employed.

Three methods of taking the ropes around curves will be seen in these sketches. In *Fig. 47* the curve has a large radius, and the tail rope is taken round a single sheave and along a narrow place, a pillar of coal supporting the roof between it and the curve. The curve of *Fig. 48* is of less radius, and no pillar is left. In *Fig. 49* is shown the plan generally adopted on very short curves; instead of taking the tail rope around a single sheave, both ropes are taken around the curve by a number of smaller sheaves.

Frequently there are extra stations on each side of the main or branch road to be worked with the same set of ropes. This can be done in different ways. If, for instance a train of cars is intended for the station (*Fig. 54*), it is taken to (*L L*), and there the ropes are knocked off; the full set stands at (*M M*), and in order to get the rope ends to this point a piece of rope the length of the train is attached to the two ends, which are then pulled by the engine opposite the ends of the full set. Thus eight connections and disconnections are necessary for each set led from the station.

A better arrangement is shown in *Fig. 55.* The ropes are knocked off while the empty set is going in at the points (*A A*), opposite to which the full set stands ready to go out. A gentle fall in the track causes the empty cars to run forward, and by the switch (*S*) they are turned into the siding (*B B*).

Fig. 56 shows still another arrangement. The middle track is the main road and the empty cars are going into the siding (*X X*) and afterwards are brought around the curve (*A*), which consists of movable rails. When the full set which stands at (*Y Y*) is taken out, the rails are removed. With this arrangement the drivers have to cross the main road every time they take the empty cars in the station, which is avoided in the foregoing plan.

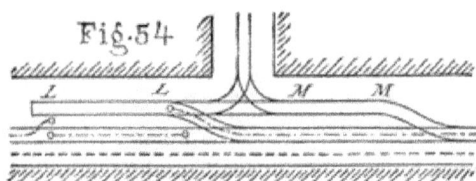

Fig. 54

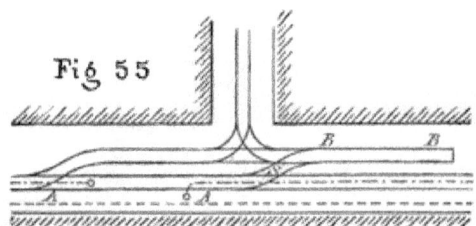

Fig 55

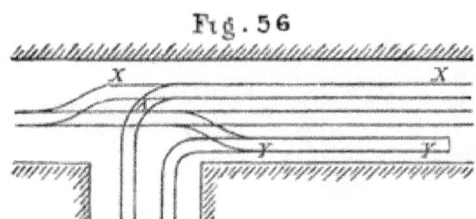

Fig. 56

An example of an extensive tail rope system, working a number of branches, is shown in *Fig. 57, Plate 13*—the plan of the North Hetton colliery in England. It consists of two separate tail rope planes, each worked by a ⅝-inch rope, and drums of four feet diameter. Plane No. 1 has a main road, with two branches on each side and a cross-cut way at the end of it. These five branches are worked by the drums marked No. 1, while the No. 2 drums work the second plane, with its branches. The ropes are shown in dotted lines. In the second west way and the cross-cut way there are two stations; the first is worked as described in *Fig. 54*, the second according to the arrangement shown in *Fig. 55*. The four curves leading from the main road to the branches each have a radius of 66 feet, while the radius of the curve in the first south way is 264 feet and of that in the cross-cut way 330 feet. No. 2 plane has one main road and three branches—two to the west and the other in a cross-cut direction. The curves to the branches have a radius of 198 feet, and the curve upon the main road 264 feet. At the far end of each of the branches there is a siding, one way for the full cars, the other for the empty cars. At the inner end of the first west way there are two "putting" stations, from which the cars are led in short sets by ponies to the siding at the end of the engine plane. The full way of the shaft siding is raised several feet to form an artificial incline called the "kep" in the English mines, and in this country known as a "knuckle." When the full cars have been drawn on this "kep," the cars are let down to the shaft as required.

One end of the axle of each set of drums is placed on a movable carriage, by means of which they are put into gear with the driving pinion. The drums are connected to the shaft by means of a clutch-gear. The engine and the drums are placed beneath the wagonway, and the wheels (W and W_1), which direct the course of the ropes for No. 2 plane, as well as several other 4-foot wheels upon these planes, are also placed under the track. The ropes for No. 2 plane come to the surface about at the point (P).

At the points (A) and (B) there are shackle-joints on both the main and tail ropes. They consist of a goose-neck socket, with a ring and clevis, as shown in *Fig. 58.* When the rope ends, to which the set is attached, are at the shaft,

Fig 58.

Shackle-Joint in main and tail-rope.
North Hetton Colliery.

these joints are always at the points (A) and (B), no matter from which way the last set came. At (C) and (C_1) the ropes are taken around the curves by small sheaves of $10\frac{1}{2}$-inch diameter, as shown in *Fig. 59;* but at most of the curves,

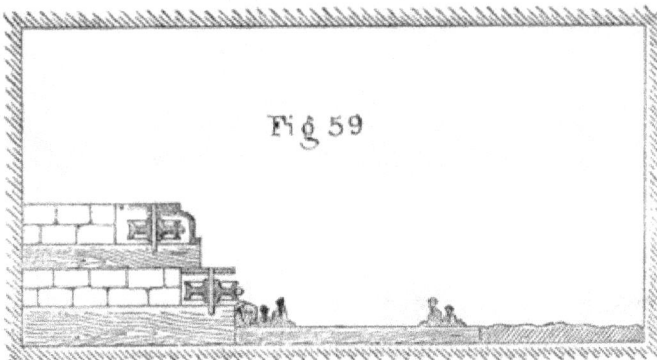

Fig 59

Arrangement of sheaves for going around
Curves in North Hetton Colliery

only one rope **follows the same, while the tail** rope passes around a single 4-foot **sheave.** This arrangement is to be preferred. The "tail sheaves" are all placed under the wagonway, and wherever the ropes have to cross the track they are arranged to pass under the rails. The total length of the main rope is 7560 feet, and of the tail rope 28,908 feet, and there are altogether 1390 small sheaves and 14 4-foot wheels upon the planes.

Method of Working the Planes.—On referring to the plan it will be seen that in No. 1 plane the ropes connected with the engine are those of the cross-cut way, and that the ends of all the other branch ropes are lying at the branch ends. It **is supposed that** a full set has just arrived at the shaft, and **that the next empty** set has to **go into** the second south way; while the rope ends at the shaft are **disconnected from the full and attached to the** empty set, the boy attending the switches at (*B*) **is disconnecting the shackles** (*S S*) and connecting them to (*T T*); this is done in about two minutes, and is generally finished before the set at the shaft is ready to come away; **the** boy then opens the switches for the second south way, and everything is ready for the set to go **in.** The empty cars **are** taken into **the** branch and the full **train returns to the** shaft before the ropes are altered again. Should the **first north way** next be ready, the ends (*E E*) are replaced by (*F F*), the switches are placed right, and the **empty set goes** in and **the** full set comes **out.** If the cross-cut way be next ready, **it will be seen** that **to** put the ropes right for this **branch** four rope ends **will** have to **be connected, two at** station (*A*) and two at (*B*).

On Plane No. 2 it will be noticed that the ropes connected with the engine **are** those of the third west way, **and** here also the set is supposed to be at the shaft. All branches **on** Plane No. 1 are dipping inward, while those of Plane **No.** 2 dip **towards the shaft.** The branch ropes on Plane No. 2 **are** connected **in the same way as** on Plane No. 1, and here also it is necessary to connect **four rope ends** when **the** third west way has to be worked, if the second and the **first west way** have been worked before it. **In the** first west way **the** grade is **found** heavy enough to **cause the outcoming full cars** to pull the tail rope after **them**; in taking the empty set inward, **the main** rope is knocked off at the point (*R*) and the train is pulled in by the **tail rope**; the full set is afterwards let down the incline by the single tail **rope to** (*R*), **where the** main **rope,** which **is** necessary to pull the set **on** the knuckle, **is attached.** **The** drum **man** sometimes brings the train out of this branch by the brake, while the engine is working another way.

The ropes are attached **to the** cars by a knock-off **link, as** shown in *Fig. 51.* A set consists of 22 to 35 cars, **and** moves with a speed **of ten** miles per hour on **an** undulating grade of which the heaviest, dipping outward, is $6\frac{1}{2}$ feet in 100 feet. The **average** duration **of** the ropes **is three** years.

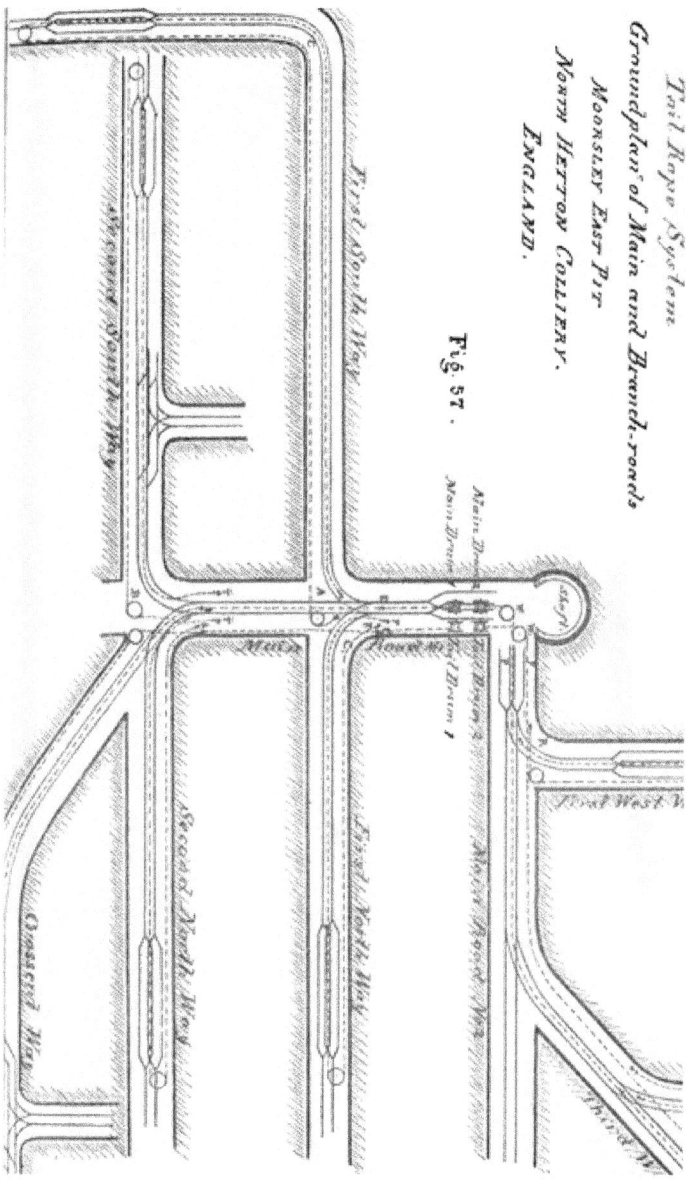

Tail Rope System.

Groundplan of Main and Branch-roads

MOORSLEY EAST PIT

NORTH HETTON COLLIERY.

ENGLAND.

Fig. 57.

IV. THE ENDLESS ROPE SYSTEM.

The principal features of this system are as follows:

1. The rope, as the name indicates, is endless.

2. Motion is given to the rope by a single wheel or drum, and friction is obtained either by a grip-wheel or by passing the rope several times around the wheel.

3. The rope must be kept constantly tight, the tension to be produced by artificial means. It is done in placing either the return-wheel or an extra tension wheel on a carriage and connecting it with a weight hanging over a pulley, or attaching it to a fixed post by a screw which occasionally can be shortened.

4. The cars are attached to the rope by a grip or clutch, which can take hold at any place and let go again, starting and stopping the train at will, without stopping the engine or the motion of the rope.

5. On a single-track road the rope works forward and backward, but on a double track it is possible to run it always in the same direction, the full cars going on one track and the empty cars on the other.

There are several mines in the Monongahela and Ohio valleys, and one in the Pennsylvania anthracite region, which have adopted this method of conveying coal, but as a rule it has not found as general an introduction as the tail rope system, probably because its efficacy is not so apparent and the opposing difficulties require greater mechanical skill and more complicated appliances. The advantages of this system are, first, that it requires one-third less rope than the tail-rope system. This advantage, however, is partially counterbalanced by the circumstance that the extra tension in the rope requires a heavier size to move the same load than when a main and tail rope are used. The second and principal advantage is that it is possible to start and stop trains at will without signaling to the engineer. On the other hand it is more difficult to work curves with the endless system, and still more so to work different branches, and the constant stretch of the rope under tension or its elongation under changes of temperature frequently causes the rope to slip on the wheel, in spite of every attention, causing delay in the transportation and injury to the rope. The pulling rope runs in the centre of the track, supported by wooden rollers, while the loose or pulled rope generally runs on the side of the road, supported by rollers either on the ground or sometimes overhead, similar to the tail rope. As the strain in the latter is considerably smaller than it is in the pulling rope, it may, like a tail rope, consist of a smaller size in cases where the rope works backward and forward. On a double track, however, where the load changes from one to the other, it must be of one size throughout.

As an example of this we mention, first, one at the State Line mine (*Fig. 60, Plate 14*). The plane is $1\frac{1}{2}$ miles long, with several slight curves and an undulating grade varying from level to 4 per cent. inclination. It is worked by a $\frac{3}{4}$-inch steel rope, made of four strands, with seven wires in each, and laid up in a very long twist. The pulling rope is supported every 12 feet on 6-inch wooden rollers, 15 inches long, with 1-inch iron axles resting on wooden bearings. The rollers for the loose rope have the same diameter, but only a length of eleven inches. They must be renewed, on an average, every twelve months, some wearing out in half a year, others lasting eighteen months or longer. For guiding the rope around the curves, smooth-faced iron rollers of 12-inch diameter, as shown in *Fig. 62*, are placed in the centre of the track at distances of six to ten feet, according to the radius of the curve. At point (*A*) the ropes make a sharp bend sideways as well as downward towards the driving-wheel, the loose rope being led around an ordinary roller as just described, the pulling rope around a 3-foot rubber-lined pulley.

The driving machinery (*Fig. 61*) consists of a double-acting engine with 12-inch cylinders by 24-inch stroke, a 24-inch pinion with 8-inch face, working a 6-foot spur-wheel, which is on the same shaft with the driving-pulley. The latter is a double-grooved, rubber-lined wheel, 6 feet 2 inches in diameter, around which the rope is wound with two half-turns for obtaining friction. The return-wheel is placed under the rails in front of the parting inside the pit. It consists of a 6-foot rubber-lined sheave, with a 3-inch axle. Before passing out on the road the rope is led from the driving-wheel around a single-grooved tightening pulley of 6-foot diameter, which, resting on a sliding frame, can be moved backward or forward by means of a 2-inch set-screw that bears against the timber frame of the building. This arrangement works well, except that in wet weather the rope slips sometimes when the engine is started too suddenly.

Referring to *Fig. 60*, the method of operation is the following: The empty cars return from the tipple on track (*L*), which has a knuckle and down grade towards (*A*), where the trains are arranged. A set consists of from 24 to 60 cars; the front car is hitched to the splicing-link of the rope with an ordinary chain, 20 feet long, having a hook at one end and a clevis at the other. The former takes hold of the splicing-chain, while the latter is attached to the pulling-bar of the car by a very tapering pin, which is secured to the side of the car by a little chain. This method of attachment is the reason for making the splice of the endless rope in an unusual manner. Instead of a regular rope splice, a piece of chain, about 15 feet long, connected to the two rope ends by clevis and pins, is used. The chain has two swivels, to allow the rope to turn, and the ends of the rope are provided with very slender sockets, 4 inches long, which are fastened by turning the wire ends over and pouring lead in all crevices (*Fig. 63ª*).

There are a few other splices in this rope, as shown in *Fig. 63ª*, which were put in because of occasional breakages.

The principal splice never goes around the end sheaves, and moves only between (*A*) and (*B*), the points where the empty and full sets are attached.

Each trip is accompanied by a train-rider on the front car. When going inward he pulls the coupling-pin at point (*B*) and the set runs by its own momentum into track (*M*) of the parting, while the man jumps down and reviews the passing cars. After the last car, called the caboose (which is a special car provided with a brake), arrives at the switch, it is a sign that everything is all right and that no car has been lost. The caboose is now attached to

Endless Rope System.

Slate-line Mine, East Palestine.

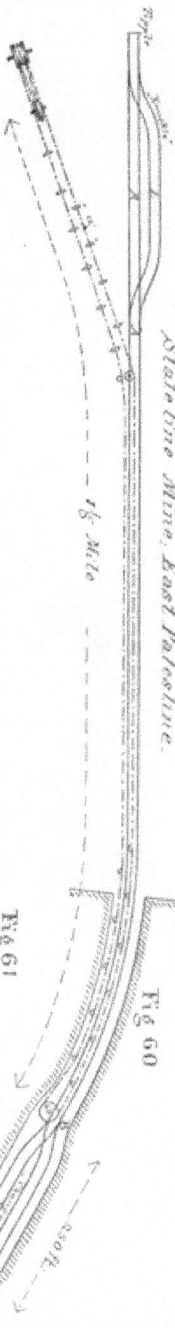

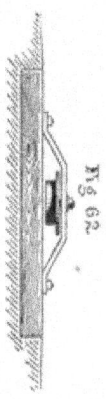

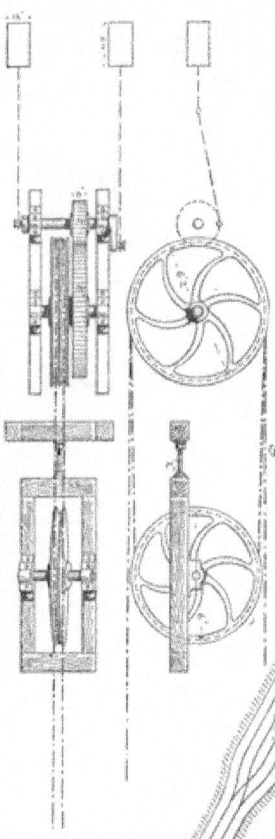

Fig. 62

Guiding Sheave in Curve.

Fig. 64

Pit Car.

Side Elevation.

Front Elevation.

Fig. 63ᵃ.

Fig. 63ᵇ.

Tension Sheave.

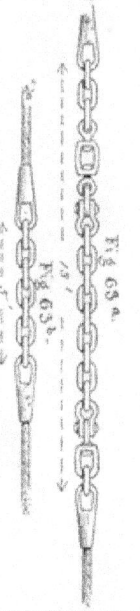

Coupling of endless rope.

Fig. 61

Fig. 60

1½ Mile

Pl. 14.

Pl. 15.

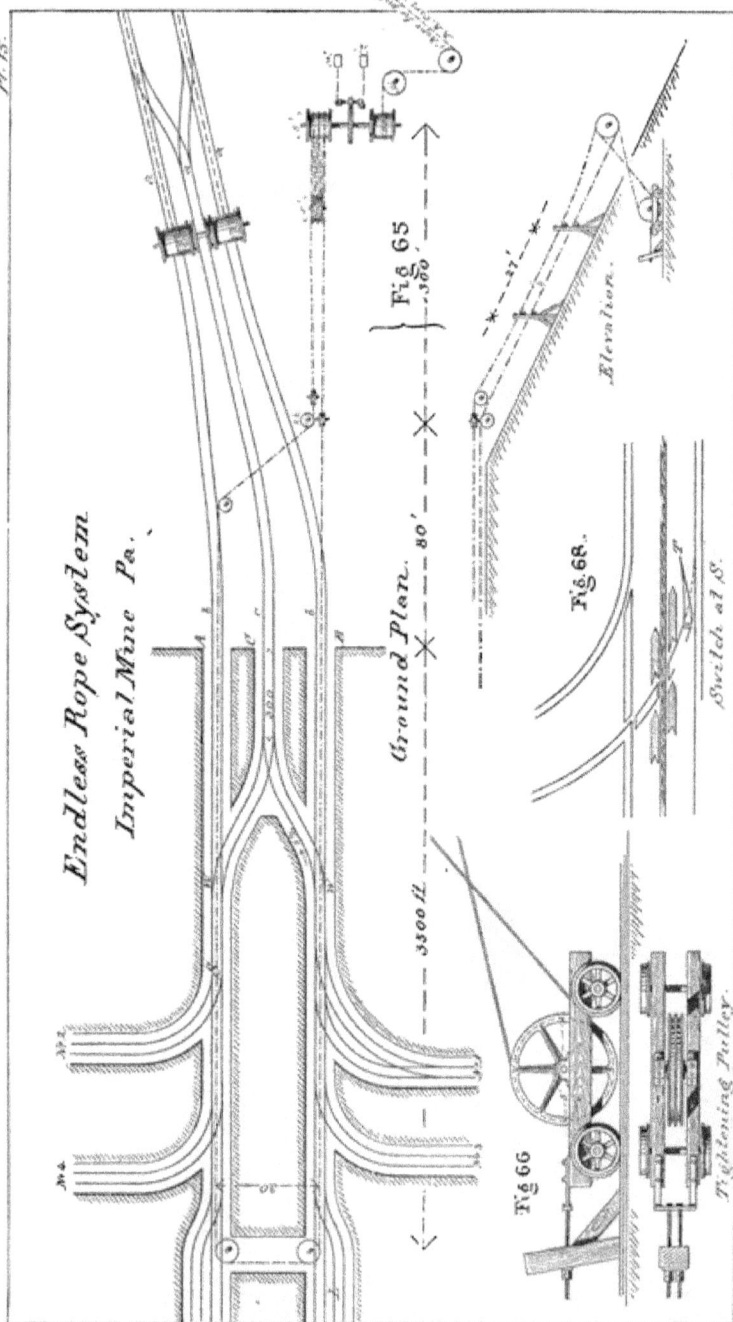

Endless Rope System.
Imperial Mine. Pa.

Ground Plan.

Fig 65

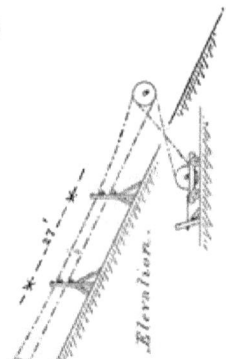

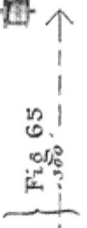

Elevation.

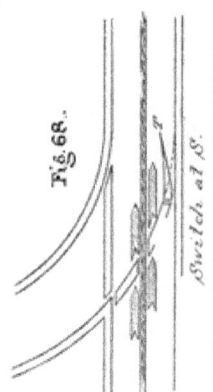

Fig. 68.

Switch at S.

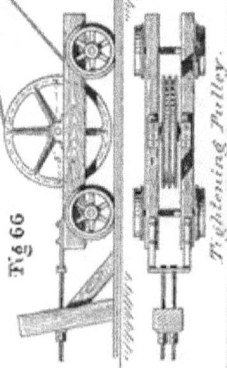

Fig 66

Tightening Pulley.

the full set standing on track (*Q*), the other end hitched to the rope, and the signal given to the engineer to start the rope. On the outward trip the trainman takes position on the caboose, and when arriving at point (*A*) he detaches the coupling-chain and the train descends on track (*R*) alone to the tipple. The cars of this mine have different proportions from those hitherto described, and for emptying the coal are provided with doors moving on side hinges (*Fig. 64*). Each car contains 2400 pounds of coal, and the daily output is from 750 to 800 tons. The average running speed is ten miles an hour.

It will be noticed from the above description that the operation in this mine does not differ materially from that of a tail rope plane. A full set of cars is arranged beyond the return-wheel, at the parting, and no stops are made between the two termini. A tail rope system, with a rope spliced endless if preferred to loose ends, similar to the Gray & Bell mine, would therefore answer equally well for the conditions of this mine, the only advantage derived from the friction system consisting in a saving of one-third the length of the rope.

The Imperial coal mine is a second example where the endless rope system has been successfully applied. Here the conditions are different. The train is not arranged at one certain point, but the single cars are collected from different stations along the main road, so that only after having passed the last cross-pit will the train have its full complement of cars. With no other system could this mode of operation be carried on as conveniently as with the endless rope. *Fig. 65, Plate 15*, represents the general arrangement in ground plan and elevation. From the pit mouths to the return-wheel the plane, which consists of a double road in two separate entries, has a length of 3500 feet, and the whole rope is 8750 feet long. It is a ⅝-inch steel rope, with seven wires to the strand, spliced endless by a regular rope splice of 30-foot length. Contrary, however, to the usual way of removing the hemp centre and putting the strand ends in its place, the latter are tucked under the other strands. This was considered an improvement, because, owing to the long twist with which the rope was manufactured, the strand ends had pulled out. Motion is given to the rope by a wooden drum of 5-foot diameter, around which it is wound with three half-turns. From this it is led around a double-grooved tightening pulley, which rests on a carriage fastened by a screw to a fixed post, and which serves as the means to keep the rope at a uniform tension (*Fig. 66*).

As will be seen in the sketch, the ropes running from one sheave to the other cross each other so as to get the largest possible bearing surface between the driving-wheel and the rope.

The question of gripping the rope and starting and stopping at will while the rope is moving has been solved in an ingenious manner by means of a special grip-car called the "dilly," invented and patented by the Imperial Coal Co. (*Fig. 67, Plate 16*). It consists of a carriage with two heavy longitudinal timbers which support the bearings of two rubber-lined sheaves of 4-foot diameter. The rope, passing with one half-turn around each of these wheels, causes them to revolve, while the carriage is stationary, but in pressing on the lever (*L*) their revolution is stopped by the brake-bands and the whole car moves along with the velocity of the rope. It is therefore in the power of the dilly-rider to start and stop at any place, and to do so without a shock, if he presses on the lever or relieves it gradually. The two wheels are placed in the bearings with a slight inclination, to allow the two parts of the rope to pass each other. Two other levers, (S) and (P), will be noticed on the sketch; the first serves to draw the coupling-pin, the second to put a brake on the carriage-wheels for stopping the "dilly" more readily when its own momentum would still keep it in motion.

From the driving-drum to within 70 feet of the pit mouth the rope is supported by 10-inch rollers, on wooden frames, but inside the pits it rests on 6-inch rollers, placed at distances of 20 feet. The driving engine is double-acting, with 12x24-inch cylinders, and a gearing of three to one. On the same shaft with the driving drum of the endless rope there is another drum of the same diameter, which works a separate engine plane with a 1-inch rope for hauling the railroad cars under the tipple, as described in a former chapter and illustrated in *Fig. 25.*

Referring to *Fig. 65* it will be noticed that about 300 feet from the pit mouth the two main entries, which are 30 feet apart, are connected by a common central entry, through which all loaded trains are taken to the check-house at (*a*), while the empty sets return in a straight line on tracks (*b*). When an outcoming train approaches the switch (*W*), the dilly-rider slackens the speed, so as to be able to draw the coupling-pin and detach the dilly from the cars. A boy, accompanying each train, jumps down and sets the switch at (*W*). The cars run alone into the middle entry on a down grade of two per cent., while the dilly goes in a straight line as far as (*A*) or (*B*). Outside the pit the middle track has a slight rise towards the head of the self-acting incline, to check the acquired speed of the cars and to carry them not farther than convenient for letting them down the incline in sets of three. At point (*a*) the three tracks are on the same level, but the two outside tracks rise from here towards the pit mouth so that point (*c*) is about seven feet lower than points (*b*). At (*W*) the three tracks join again in the same level. The empty cars coming up the gravity incline are coupled together in sets of 40 on track (*b*), and after the dilly has been attached the signal is given to the engineer to reverse the engine, this at the same time being the signal to the dilly-rider at the parting (*L*) to get ready for going out. Suppose three cars were to be left at cross-entry No. 2, the train-boy would ride in the car ahead of the three and would cast them off when approaching the siding, the dilly-rider slackening the speed a little for this purpose. For entering the side entry the cars necessarily must pass over the rope. The latter, therefore, as shown in *Fig. 68*, is sunk into a groove and by four wooden guiding blocks prevented from leaving the same, while the rails are raised above the ordinary level to avoid all danger of cutting the rope with the flanges of the wheels. After the train has passed the switch the train-boy closes the tongue (*T*) and the three detached cars run alone into the side entry, the whole operation being done without stopping. At every cross-entry where cars are to be left the same manipulation is repeated.

The outward trip is conducted in the following manner: Starting with a few

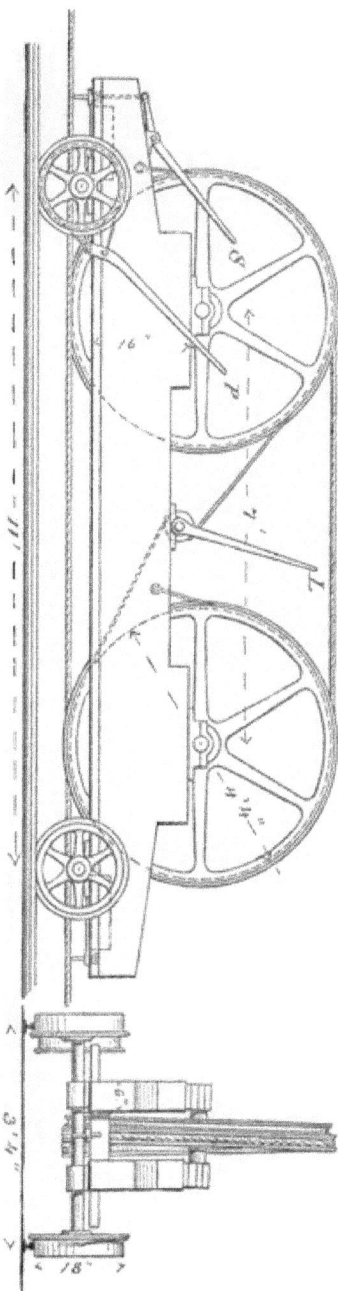

Fig. 67.

Patent Grip Car "The Dilly"
of the
Imperial Coal Co.

Pl. 16.

Pl IV.

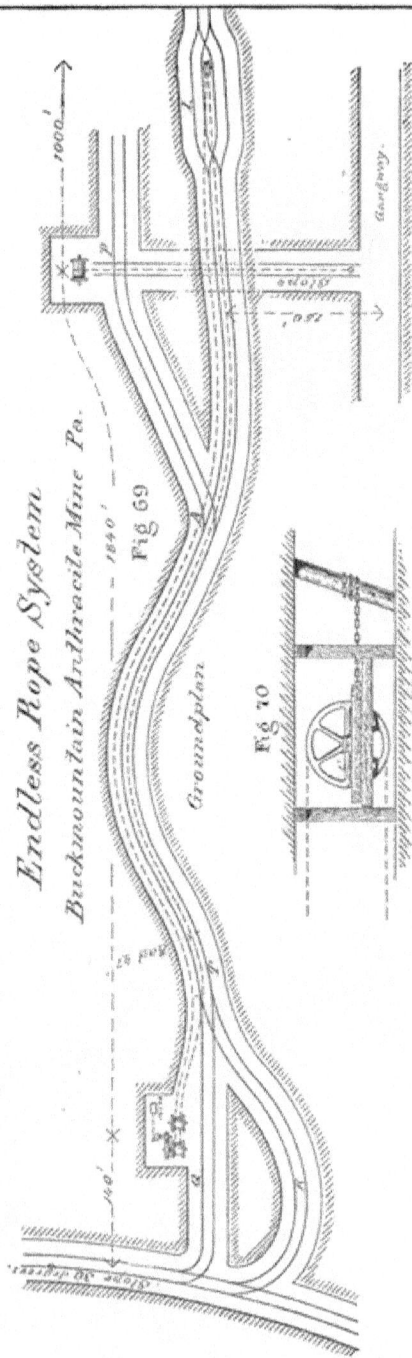
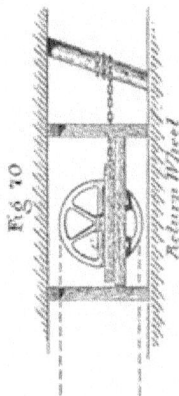
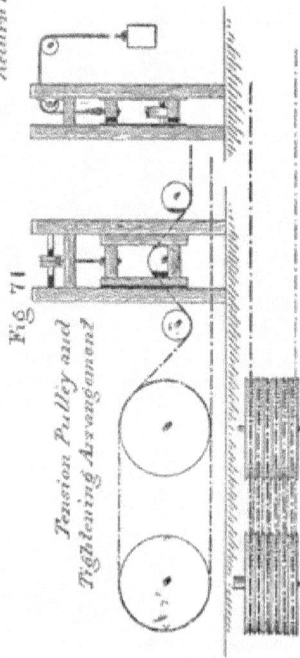
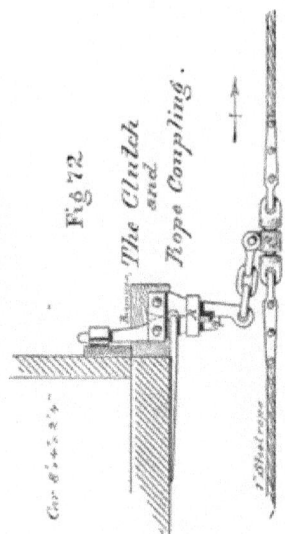

Endless Rope System.

Buckmountain Anthracite Mine Pa.

Fig 69

Groundplan

Fig 70

Return Wheel

Fig 71

Tension Pulley and
Tightening Arrangement

Fig 72

The Clutch
and
Rope Coupling.

Car 8'×4'×2½"

Gangway.

cars from the parting (L), the dilly stops at the first cross-entry and the train-boy attaches a chain about 15 feet long to the full cars standing on a side track, which are drawn by the grip-car on the main road. By slacking the speed these cars catch up with the train and are attached with the regular coupling by the train-boy, who then takes the chain and jumps in the last car to repeat the same operation at the other side entries.

The switch tongue (T) is self-acting for all outgoing cars : if shut the dilly will open it, and if open the cars from the side tracks will close it. On the inward trip, however, the dilly-rider can open the tongue by means of a long switch-rod before the dilly reaches there.

The regular running speed is about six miles per hour, and the daily output of coal amounts from 1200 to 1400 tons.

The endless rope system of the Buck Mountain anthracite mine is illustrated in Fig. 69, Plate 17. The driving engine is situated in the pit, about 140 feet from the foot of a slope of 30 degrees inclination, which is worked by a separate rope and engine. The length of the plane from drum to return-wheel is 2840 feet ; it is nearly level, falling, in the whole distance, only three feet, but has several curves, the smallest having a radius of 67 feet. The working rope is of steel, of 1-inch diameter, and lasts about 12 or 13 months. It is driven by a single-acting engine, with a 14-inch cylinder by 48-inch stroke. Friction between the driving-drum and the rope is obtained by giving the latter six half-turns around this and another grooved drum of 7-foot diameter. Tension is given to the rope in two ways : first, the return-wheel, a 6-foot rubber-lined sheave, is placed on a sliding frame attached to a fixed post by a chain, which may be shortened to take up the slack of the rope (Fig. 70). A second method, shown in Fig. 71, is self-acting. It consists of two fixed and one movable sheave, the "loose" rope passing under the former and over the latter. This rests in a frame which slides between two upright posts and is connected by a light rope with a counter-weight that keeps the rope constantly in equal tension. Though effective in its results, this method is not desirable, because the sharp bend around the small sheaves is very injurious to the rope.

The clutch with which the cars are attached to the rope is shown in Fig. 72. Two hinged pieces of iron, after being hooked into the eye of the pulling-bar, are kept closed by a ring (R) and a cotter (q). These can be knocked off while there is tension in the rope, and the clutch will then unhook itself, but the rope must be stopped for attaching as well as for detaching. At the upper end of the clutch there is a cross-piece of wood, which rests on the bumpers of the car, while with the lower end it is fastened by a clevis to a ring (s), which moves freely on the link of the rope splice.

In looking at the plan (Fig. 69) it will be noticed that 1800 feet from the parting a sloped road crosses the main road, led under the same, on which by a separate rope and engine at the head of the slope the coal from another gangway is taken to point (P) and afterwards lowered to (A) to join the train on the main track.

The mode of operation is therefore as follows : Starting from the parting (L) with eight or ten cars, a stop is made at point (A) by a bell signal to the engineer, and four or five cars, which were let down from (P) by a brake, are attached. When the train arrives at (T) the engine stops. The full cars are taken by mules on track (Q) to the foot of the slope, while the empty cars coming from the same run on a down grade, on track (R), to point (T), where the ingoing set is arranged. The cars for the lower gangway are taken from (A) to (P) by the same rope which works the slope, while the remaining cars go to the parting (L). A train consists of 13 to 22 cars, each of them containing 4200 pounds of coal. The average time for a round trip is from 15 to 18 minutes. The ropes are supported on rollers placed at distances varying from 6 to 100 feet. Exclusive of wear and tear of the rope, the total cost of transporting the coal about 3000 feet is estimated at one and a half cents per ton.

In England the endless rope system has found considerable favor, and is in use at the Shire Oaks and Cinderhill collieries in Nottinghamshire, the Newsham colliery in Northumberland, and the Eston mines in Yorkshire. The methods

Fig 73

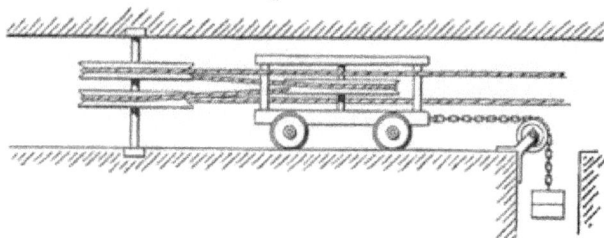

Fig. 74

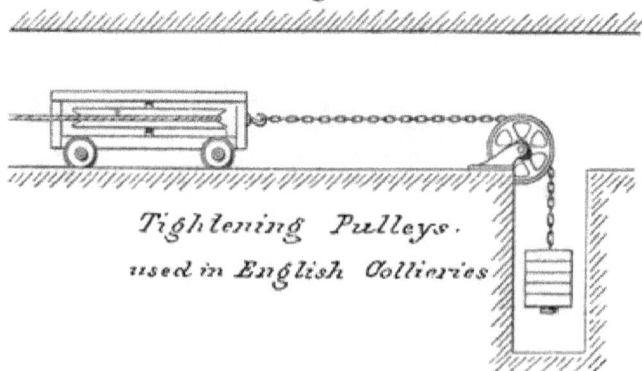

Tightening Pulleys.
used in English Collieries

applied for tightening the rope in these collieries consist generally in placing either the return-wheel or the tightening pulley on a carriage to which a hanging weight is attached (Figs. 73, 74).

Pl. 18.

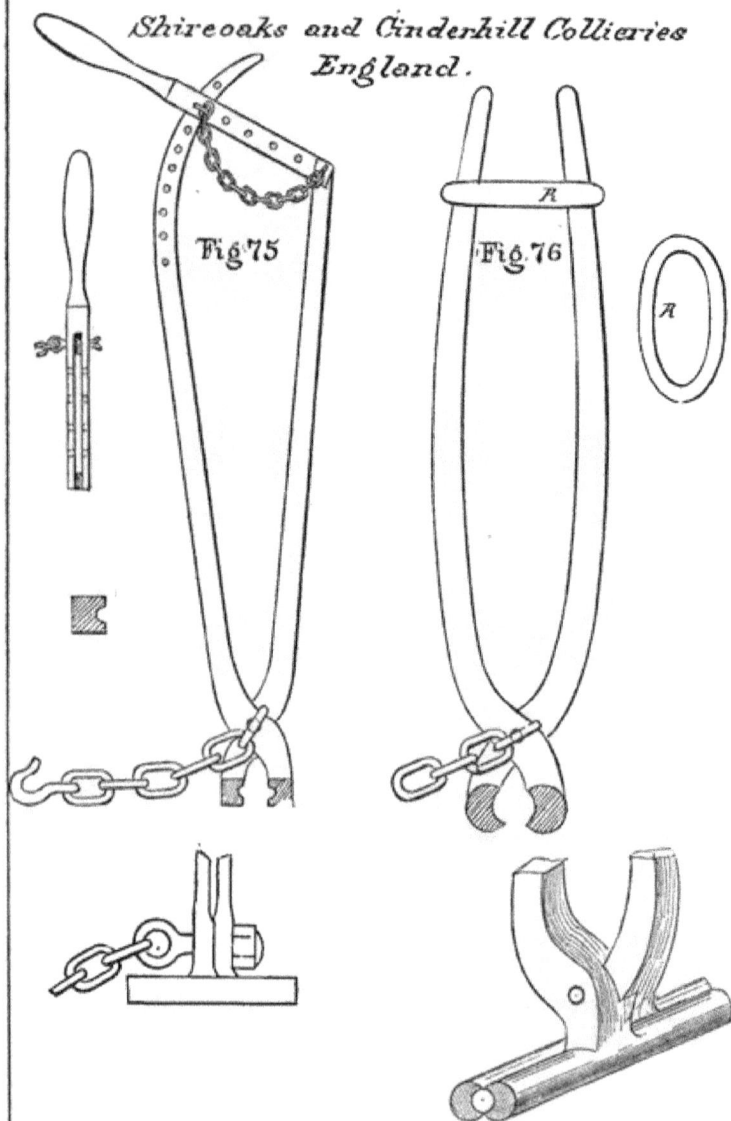

Rope-Clutches
used at
Shireoaks and Cinderhill Collieries
England.

Fig.75

Fig.76

A

A

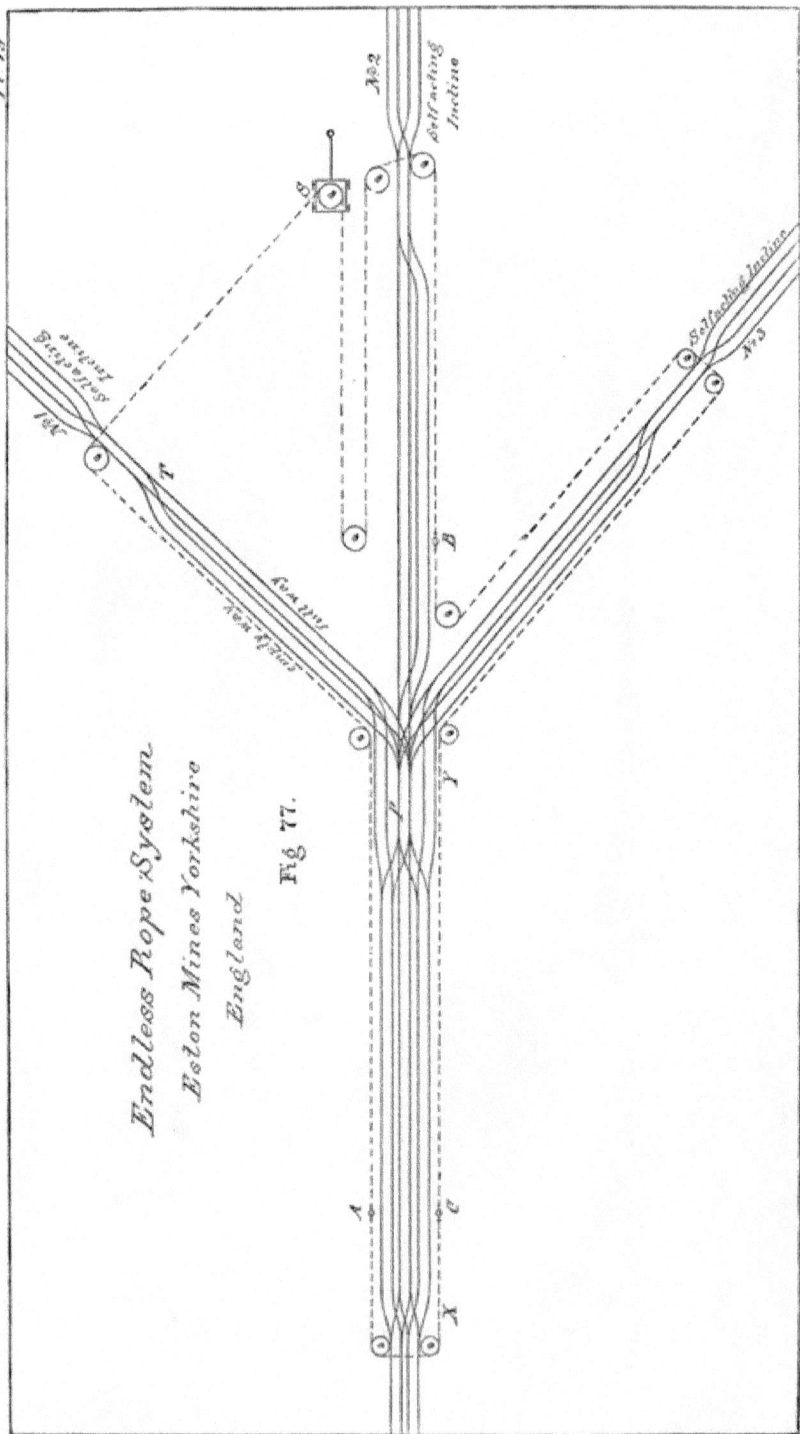

Pl. 19.

Endless Rope System.
Eston Mines Yorkshire
England

Fig 77.

No 2
Self-acting Incline

Self-acting Incline
No 3

Self-acting Incline
No 1

Full way

Full way

Empty way

T

B

X

C

A

Y

Friction for driving the rope is frequently supplied by a grip-wheel, as was illustrated in *Fig. 6, p. 8*. The clamps or grips for attaching the cars to the rope, used at Shire Oaks and Cinderhill mine, are of the pattern shown in *Figs. 75, 76, Plate 18*. They are applied while the rope is in motion—in the first-named mine by a boy riding in the front car, and in the latter mine by a man traveling on foot alongside the train, which moves only at a rate of $2\frac{1}{2}$ miles per hour. It has been found that there is no difficulty in passing these clamps around the curves.

The plan and rope arrangement of the Eston mine, in Yorkshire, is shown in *Fig. 77, Plate 19*. It is distinguished from other endless rope systems in that three branches and one main road are worked by one rope.

From the bottom of three self-acting inclines run three branches, joining at one point a fourth branch which leads to the top of an incline. The endless rope is used to convey the full cars from the three sidings to the main road, and to bring empty cars back again. The rope, shown in the plan by a dotted line, is one inch in diameter, and kept tight by a hanging weight attached to sheave (S), which is placed on a tram. At the points (A), (B), (C) there are links or sockets in the rope, by means of which the connection between the rope and the sets of full and empty cars is made. Each of the links has a certain position, to which it is always brought by the engine after having been used in moving a set of cars. The link (A) is used for drawing the empty cars to the siding of No. 1 incline and the full cars from both No. 1 and No. 2 siding to the main road ($X\,Y$). The link (B) draws the empty wagons to No. 2 siding, and the link (C) to No. 3 incline. When an empty set is required for No. 2 incline, the link (A) brings the set to the points at (P), from where it is conveyed by the link (B). The set consists of twenty cars. Suppose a set of empty cars had to be taken to the siding of No. 1 incline: By means of a short 2-inch steel chain, with a hook at each end, the set is attached to the rope and drawn upon the siding to point (T), where the rope is disconnected. Should a full set be ready for removal to ($X\,Y$), the rope is attached and it is brought back again to its original position. If no wagons are ready, the engine draws the rope back empty. The rope is also frequently used in bringing the full cars on the sidings when they fail to run far enough on the self-acting inclines.

A certain method of the endless rope system, employed in England in cases of a double track, is well worthy of consideration. It is in use at the Bridge, Meadow, Moor and Scotlane pits and the Mesnes colliery, and has several advantages. The rope, in place of being supported near the ground on rollers, rests on top of the cars, saving the expense of rollers and avoiding the friction on the rope. The cars are attached singly or in sets of two to twelve at regular distances, the former method being preferable because it gives more frequent support to the rope. Motion is given to the rope in the usual way by driving-wheels, around which several turns are taken to secure friction. The motion is slow, only from 0.8 to 1.2 miles per hour, but as the service is continuous it is possible, even at this slow speed, to convey more coal to the shaft than the latter generally can accommodate. Curves and branches can also be worked by this system.

Fig. 78, Plate 20, represents the plan of the Bridge pit, near **Wigan**. All planes are laid with a double track—one for the ingoing empty cars, the other for the outgoing full cars. The general width of the pits is 10 feet **4 inches**, and all roads rise towards the shaft at a rate varying from one foot in eighteen feet to one foot in sixty-two feet, with exception of the slant-way, which falls towards the shaft with a decline of one foot in forty-seven feet. The total length of the main driving-rope, which is of steel of $1\frac{1}{4}$-inch diameter, is 2361 feet. It passes two and a half times around the driving-wheel, which has a diameter of 14 feet, and the outgoing rope goes past the shaft to the tightening-pulley (**A**). This pulley, of 9-foot diameter, is placed on a strong tram and the rope is tightened by means of a screw attached to the carriage, the other end being secured to a piece of fixed timber. There are two sets of pulleys near the shaft; one (**B**) for working the main way and the slant way connected with it, and another (**C**) for working the chain brow way. The driving-rope, in coming from the tightening-pulley, passes once around the pulley (**C**) and twice around (**B**), there being a much greater load on the plane worked by the latter pulley. Both pulleys have a diameter of 9 feet. The tightening-pulleys have a diameter of only 5 feet. On shaft (**B**) and (**C**) are two smaller sheaves of 6 and 5 feet diameter, arranged so as to be put in and out of gear. The rope working the main road is of **steel**, one inch in diameter, and that on the chain brow way has the same diameter, but is made of iron.

The main road rope passes around a tightening-pulley (**D**), about 2070 feet from (**B**), and then it passes twice around the pulley (**E**). There are two pulleys on this shaft, the one for the driving-rope being **6-foot** diameter, and the one working the slant **way** 5-foot. The rope on the latter is a 1-inch iron rope, and is tightened by the apparatus at (**F**).

The chain brow way has a curve at (**H**), around which the rope is taken by means of two $4\frac{1}{2}$-foot pulleys, placed horizontally. This curve is at an angle of 118 degrees, and the rails are laid at a radius of only 15 feet. The rope passes round the **tail-sheave** (**J**), which also acts as tightening-pulley. The cars are attached to the rope either singly or in sets of two, at a distance of about 60 feet

Fig. 79

Attachment of car to overhead endless rope
at Bridge Pit England.

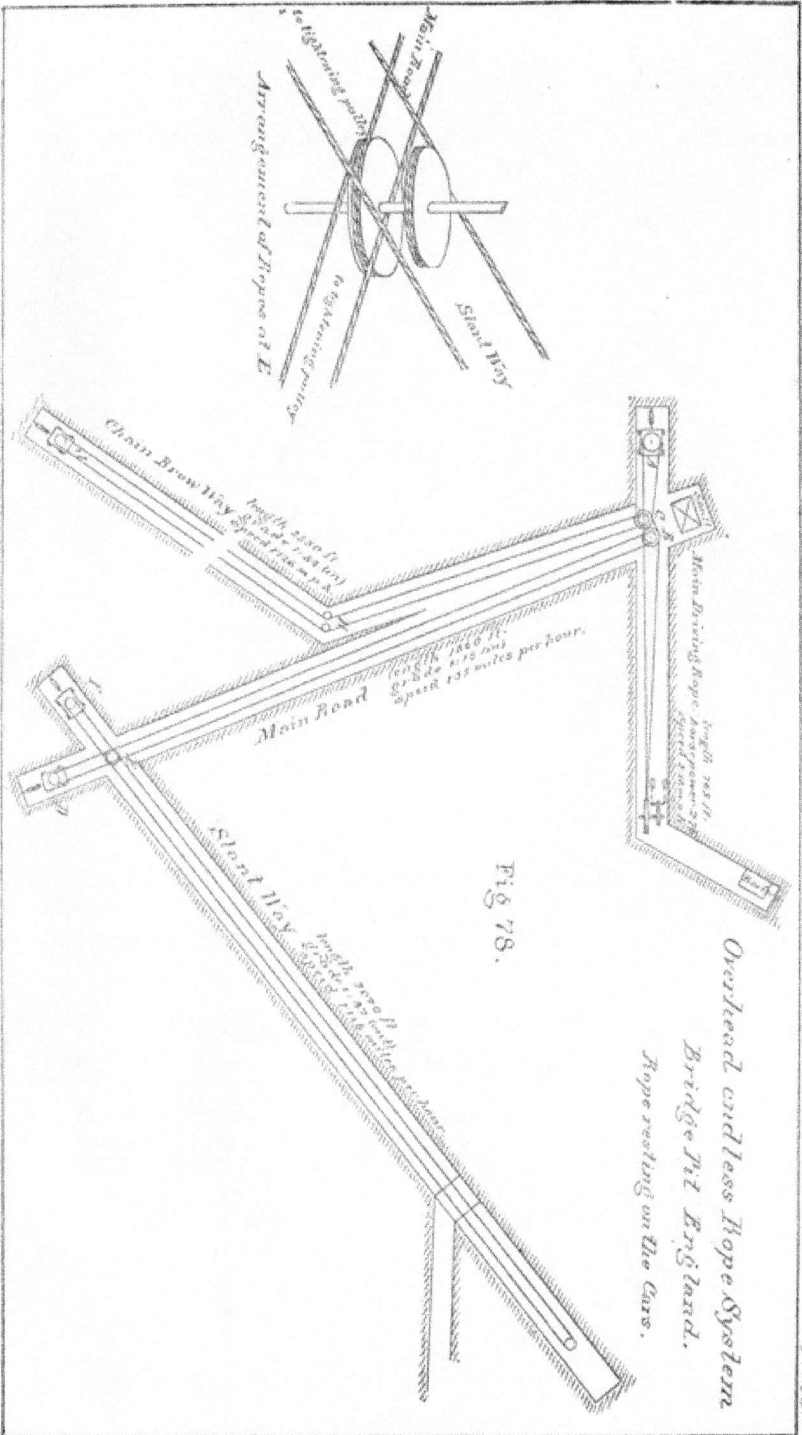

Pl. 20

Overhead endless Rope System
Bridge Pit England.
Rope resting on the Cars.

Fig 78.

Arrangement of Ropes at E.

Main Rope

tightening pulley

tightening pulley

Slant Way

Chain Brow Way

Main Driving Rope

Main Road   length 1860 ft.
Speed 135 miles per hour.

Slant Way

from each other. For making this connection a ⅜-inch iron chain, 6 feet long, with a hook at each end, is used. When the road has a regular grade either way, one chain is sufficient at the end of the car going to rise, but on level or undulating roads two chains are necessary. *Fig. 79* shows the method of hitching the chain. After it has been hooked with one end in the coupling-bar of the car, the other end is passed twice over the rope, the hand being introduced under the rope to let the chain slide loosely on the moving rope till the hook is secured. When the right number of coils of chain have passed over the rope, the hand is withdrawn, the point (*H*) is brought over the hook, and the chain is pulled tight. An expert hitcher can do it quick enough, before the rope has time to move on, and does not need to introduce his hand between the coils. For disconnecting the chain at the fore end, when it is tight, it is necessary to put the foot on it and press it down to make it loose enough for disconnection.

The usual time for attaching both chains is about twelve seconds, and a little more for disconnecting them. The chain rarely slips, but when it does, or breaks, the damage done is generally heavy. The slow speed of 1.1 to 1.3 miles per hour, at which the cars move, is necessary to prevent such accidents.

In the working of the endless rope the apparatus for putting the driving-wheel in and out of gear is found to be indispensable; such an apparatus is used at both the driving-wheels (*B*) and (*C*), and at (*E*) at the bottom of the main road. Thus the four branches can be worked separately. The lower driving-wheel at the points (*B*), (*C*) and (*E*) is fixed and the upper loose on the shaft, being put in gear by a catch-box worked by a lever; the pulleys can be put in gear while the driving-wheel is in motion, but the engine is usually stopped in taking them out of gear.

There are two curves on these planes—one at the bottom of the main road, worked by disconnecting and reconnecting the cars; the other on the chain brow way, which is self-acting. At the former, which turns at an angle of 72 degrees, the motion is transmitted from one pulley to another on the same shaft, as shown in *Fig. 80*. The road is laid around the curve at such an inclination that the

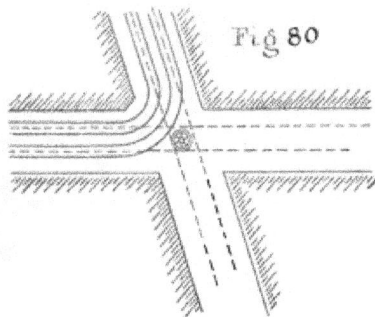

Fig 80

full and empty cars, when disconnected, run by themselves to the place, where they are again attached to the rope. This operation requires five hands—one man and four boys.

On the chain brow way the rope is taken around the curve by two 4½-foot pulleys, each inclining slightly towards the "coming-on" side. The road for the full cars is laid nearly level, and for the empty cars with a slight rise from the shaft. The pulleys are made with a large flange on the lower side, to prevent the rope slipping off and to allow the knot of the chain to pass easily in the groove of the wheel (*Fig. 81*).

Fig 81

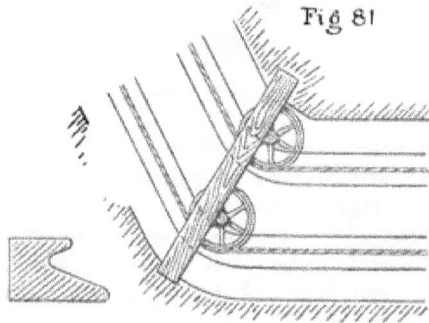

The main driving-rope wears out first, and lasts about two years; the other ropes are worn in two ways: first, by the friction of the coils of rope upon the pulleys, and, secondly, by the moving of the rope upon the cars. They last about seven years.

The method of working this system of endless ropes, as practiced in the other pits named, is about the same. At the Scotlane pit they have a different way of connecting the cars to the rope. Instead of chains passing around the rope, strong loops of hemp are fastened to the rope by a wrapping of yarn at regular distances, the hook of the hitching-chain being attached to it as shown in the sketch below (*Fig. 82*).

Fig 82

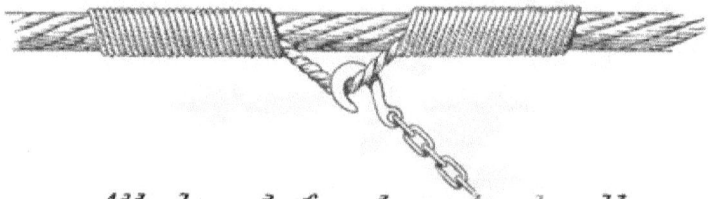

Attachment of car to overhead endless rope used at Scotlane Pit England.

The loops are made of hemp of 1-inch diameter, and last about four months; they are strong enough to draw twelve cars on a heavy grade. They are fixed to the rope, 51 feet apart, thus making a regular supply of full and empty cars necessary. Much less labor is required for connecting and disconnecting by this arrangement, but it would hardly be applicable on an irregular plane, where two loops would have to be provided for each car or set of cars. Although the rope passes one and a half times round the driving-wheel, the loops go round without causing any inconvenience.

The amount of artificial tension to which an endless rope must be subjected to prevent its slipping on the driving-wheel depends on the number of turns on the wheel, and on the condition of the surface of rope and wheel, or on the co-efficient of friction. The smaller the co-efficient of friction is and the less turns there are on the wheel, the larger must be the tension.

Calling W the total resistance, consisting of the direct weight and the friction produced by cars and rope, n the number of half-turns of the rope on the driving-wheel, Z the total artificial tension or weight attached to the tightening-pulley, f the co-efficient of friction, Q and q the tension in the pulled and loose rope, the following relations must exist to prevent slipping:

$$Q = q\, e^{f\, n\, \pi.} \qquad Z = Q + q. \qquad W = Q - q.$$

From these three equations the three unknown quantities Q, q and Z can be calculated. If W is equal to 2000 pounds, and assuming for f an average value. we find for Z, when the rope has from one to six half-turns:

$N =$	1	2	3	4	5	6
$Z =$ lbs.	5992	3330	2565	2220	2125	2030

For greater safety it is advisable in practice to increase these figures by about twenty per cent. to guard against contingencies like wet, icy weather, an oily rope, &c., because a larger direct tension is not so injurious to the wear of the rope as its slipping on the wheel. We recommend, therefore. the following corrected table:

$n =$	1	2	3	4	5	6
$Z =$ lbs.	7191	3996	3078	2724	2550	2436

At a state of rest this tension will distribute itself equally on the two ropes, but in a state of motion the tension in the pulling-rope near the driving-wheel, compared with the tension in the loose rope, will be as much greater as the total resistance amounts to, which in our example is 2000 pounds · hence the maximum and minimum tension in the rope will be:

$u =$	1	2	3	4	5	6
$Q =$ lbs.	4595	2998	2539	2362	2275	2218
$q =$ lbs.	596	998	539	362	275	218

The increase in tension in the endless rope, compared with the main rope of the tail system, where of course one ton resistance produces only one ton of strain, is therefore:

N	1	2	3	4	5	6
Increase in tension in endless rope compared with direct strain............	Per cent. 130	Per cent. 50.	Per cent. 27	Per cent. 18	Per cent. 14	Per cent. 11

In calculating the strength of an endless rope, under different conditions of the road, according to length and inclination, the results must be increased according to the number of turns given to the rope on the driving-wheel. For instance, at the Imperial mine n is equal 3, hence the rope must be 27 per cent. stronger than a main and tail rope would have to be. With the endless system, one-third, or 33 per cent. was saved in length, but considering the increased size of rope the real saving in weight or money cannot be placed higher than six or seven per cent.

Comparing the tail rope and endless systems, with their respective advantages, we may say:

1. The endless rope system is a little cheaper than the tail rope.

2. The duration of ropes is about the same, but more in favor of the tail rope system.

3. The condition of grades in the road causes no inconvenience to either of the two systems.

4. Where frequent stoppages are made, to take and leave cars at side stations, the endless system, with a good grip arrangement, is admirably adapted and undoubtedly to be preferred.

5. The cost of labor for making connections and disconnections is about the same in both systems, although in the endless system No. 2, as lastly described, it is somewhat larger.

6. Curves are not convenient in the endless system, but are easily worked with the tail rope system.

7. For working a number of branches by rope haulage, the tail rope system is certainly preferable.

FEEDING COKE OVENS BY MEANS OF WIRE ROPE.

The manufacture of coke is an important branch of coal mining, and is an industry which in the last ten years has assumed immense proportions in districts where the coal is too soft and bituminous to be directly merchantable. Along the upper Monongahela and Youghiogheny rivers, and in their side valleys, whole villages have sprung up, the inhabitants of which are engaged in the coke business, and thousands of coke ovens, extending in rows often for miles, make the surrounding country dark with their dense smoke and illuminate the night with numberless bright flames.

A coke oven is cylindrical in shape, having a spherical cupola. It is about 12 feet in diameter and 8 feet high, and occupies a length of 14 feet in the row. At the top there is a hole 15 inches in diameter for charging, and at the side a larger opening for emptying. During the process of burning the latter is kept closed with fire-brick and clay, so as to keep out the air. The burning requires 48 hours, during which time all gas is consumed and the pure carbon is left. The fire is extinguished by water, the coke taken out and the oven charged immediately anew, the remaining heat being sufficient to ignite the coal. The ovens are built in rows, numbering from 50 to 500, often in a single line but more frequently in a double line, so that on each side of the row the coke can be taken out. A railroad track runs along the foot of the ovens for the convenience of loading the coke directly from the ovens into cars. The filling is done by a specially-constructed funnel-shaped car called "the larry," which runs on a track directly over the upper openings on a single row of ovens, or between them on a double row. The larry is accordingly of different construction—for the first one with a drop-door at the bottom, and for the second with side doors and a chute for guiding the coal into the holes. Each larry contains four pit-car loads, or about four tons of coal, which is the usual quantity for charging one oven.

The larries are generally drawn by mules, walking on top of the rows between the openings from which emerge the flames of the ovens. It is a very toilsome and dangerous work, frequently injuring the animals and causing loss to the owners. Economy, as well as humane feelings for the animals, would therefore recommend replacing the mules by some mechanical power.

The use of the locomotive for feeding coke ovens is in most cases too expensive, and is only practicable in the very largest works, but a small wire rope offers a simple and cheap means for hauling the larries to and from the ovens. The engine working the hoisting apparatus of the shaft or slope of the mine will work this rope so that the whole outlay, besides the rope, would consist in the cost of a drum and a number of supporting-rollers. Some works in the Pennsylvania coke region have adopted this method with perfect success.

The arrangement of the systems at the works of the Stewart Iron Co. is illustrated in *Fig. 83ᵃ, 83ᵇ, Plate 21*, showing ground plan and elevation. There are two lines of ovens, about 560 feet long, one consisting of a single row, the other of a double row. They are built upon the principle of an engine plane, with a slight inclination, so that the full larries descend by their own weight and the empty ones are hauled back by the rope. Each row is worked by an independent $\frac{9}{16}$-inch steel rope, 700 feet long, coiling and uncoiling itself on drum (A) or (B), which form a part of the hoisting engine for the slope of the mine. (Compare, also, Fig. 26, Plate 5.) On the double row of ovens the rope runs in the centre of the track, but on the single row it runs outside, being hitched to the larry as shown in Fig. 84.

Two larries are employed for doing the work, and as 60 loads are necessary per day to fill the ovens, each larry must make about 30 trips. The engineer in charge of the hoisting engine also tends to the drums for the larry ropes. It would require the service of three mules to do the same work, and a simple calculation will prove the economy of the rope system. The maintenance of three mules per month is $45, the cost of 1400 feet of rope $140; hence in less than four months the maintenance of the mules pays for the rope, which will last from 18 to 24 months. The cost of three mules also is higher than the two drums and supporting-rollers.

Another example is illustrated by Fig. 85, Plate 21, representing the coke ovens of the Leisenring works. The endless rope system has been applied here. The splice consists of a chain 30 feet long, which is attached to the rope ends with sockets. The rope starts with half a turn from drum (D), passes around the tension sheave (S), then over the drum (C), and is guided into the line of the ovens by the sheaves (L), (W) and (Q). The tension sheave (S) rests on a sliding frame, which is attached to a fixed post by a screw. There are two rows of ovens, 100 feet apart, the rope running out on one row and returning on the other, being supported upon wooden frames in the intervening space. Both lines of ovens are in double rows, hence the rope travels in the centre of the track, resting on wooden rollers placed 25 feet apart. The attachment of the larry to the rope is shown in Fig. 86. A hook (H), moved forward and backward by means of the lever (P), slides in a slot in the bumpers, one time connecting the two timbers, another leaving a space (J) between them. The larry man lifts with his hand the splice-chain between the two timbers and catches it with the hook. If he wants to stop, he pulls the lever and places the hook in the position indicated by dotted lines. This opens the space (J), and the chain drops. At the same time he pulls the wire of a bell signal for the engineer to stop the engine, because the attachment can only be made to the splicing-chain, which for this purpose should remain under the larry. The great length of the chain is an advantage, because it allows a certain motion of the rope before the splice-chain is drawn beyond the clutch.

The larries are filled at the tipple (T), and it is customary to run four larries to a trip, the first one only being provided with the clutch, the others coupled to it and to each other by a wooden bar. In the manner described it is possible to stop at any place and to go on again in either direction, according to the signal given the engineer. After the larries have returned from their trips on one row of ovens they are switched over to the other, the rope traveling in the meantime all round the upper end until the coupling-chain arrives at (Z); the object of

Wire Rope Plane for feeding Cokeovens.

Stewart Iron Co. Uniontown.

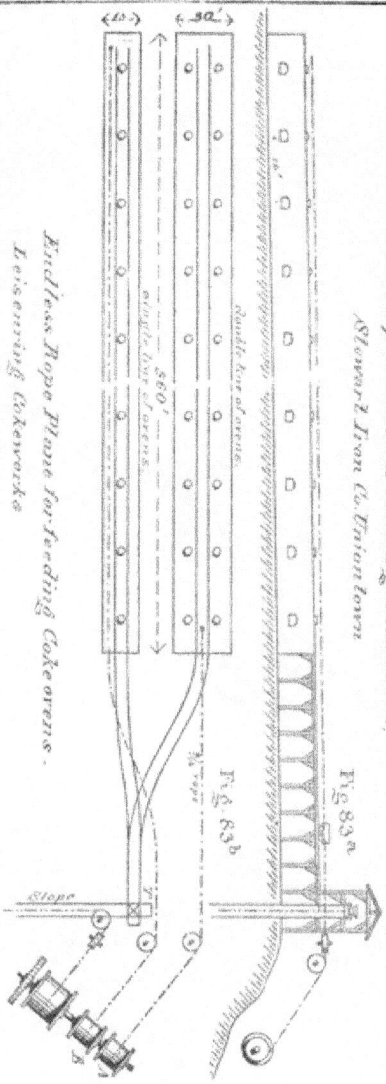

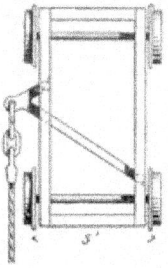

Fig. 83a

Fig. 83b

Single line of ovens.

Double line of ovens.

560'
single line of ovens.

Slope

Fig. 84.

Attachment of rope to Larry
or single line of ovens.

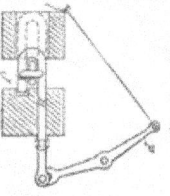

Endless Rope Plane for feeding Cokeovens.

Leisenring Cokeworks.

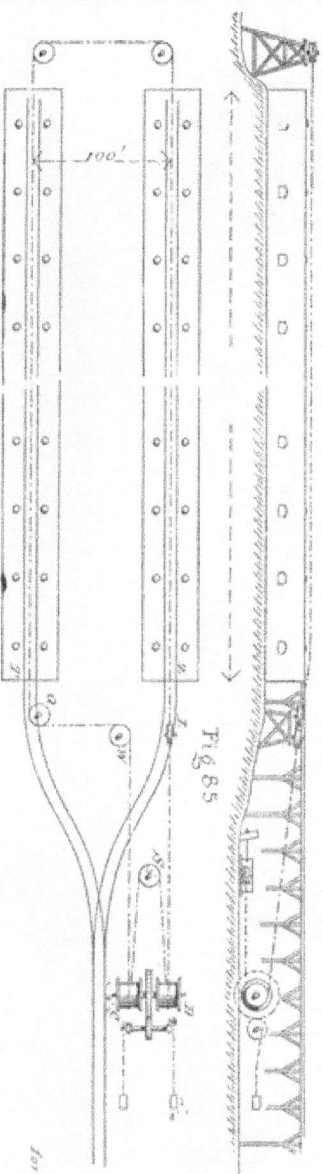

100'

Fig. 85

Rope Child.
for connecting Larry to rope

Fig. 86

Pl. 21.

this being to avoid having the **sockets and** chain go over the drums and tightening-sheave. The service is arranged so that only one shift is necessary per day, **the ovens** of the left row generally being charged in the morning, those of the **right row** in the afternoon. The rope is of steel of $\frac{5}{8}$-inch diameter, and lasts about eight months. But even with this **short** life of the rope, and in spite of a separate engine and engineer for **working the** rope, the owners find it more economical than **the service of mules—four of** which, at least, would be necessary for performing **the work.**

A more **convenient method of working would be to** provide each track with two ropes, **running between the rails in** opposite directions, there being sufficient **room** on the road for two ropes side by side, as the gauge of the larry track varies **from** five **to seven** feet. The rope would start from a common drum or grip-wheel, and, as **shown in** *Fig. 87*, could be guided in the line of one or several rows of ovens. **At the** end of its course it would pass round a wheel (*R*) **and** return alongside the first rope until joining it again at the wheel (*A*).

Fig 87

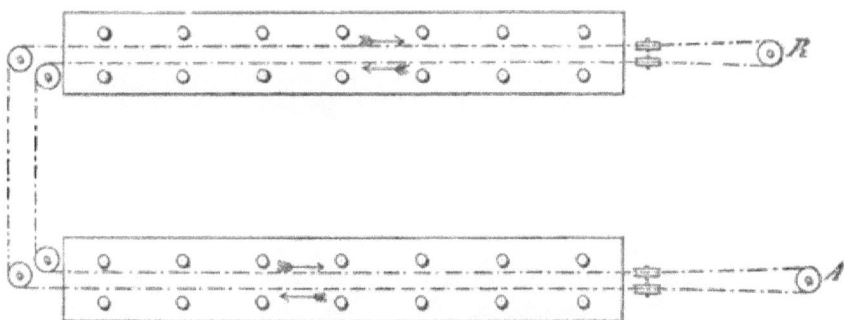

*Proposed Endless Wire Rope Plane
for feeding Coke ovens.*

This wheel could be driven **by the** mine engine, and the only manual labor necessary would be to put **it in or out** of gear according to demand. During the working hours the rope would be continually in motion, and **for** stopping and starting **the** larries at will **it must** be possible to **grip** the **rope** at any point. The clutch of the Leisenring larries would therefore **not answer, but** one of **the** clutches shown in *Figs. 75, 76, Plate 18*, could **certainly be used** with advantage. **They are** simple, and their efficiency has **been** proved in practice. **They can be applied to any** larry and do **not** require that the rope should be lifted **up, as they slide easily** over the rollers. **To** facilitate this sliding **we would** suggest, as an improvement, the tapering of **the** ends of the jaw, instead of leaving them, as shown in the **sketches, of** the same thickness throughout. The short chain with which **the clutch** is attached to the larry will enable the **larryman to** grip either of the two ropes, according to the direction in which he is bound. **For the best preservation of the rope the slowest possible speed is advisable.** Take, for instance, the case of Stewart Iron Co., who require 30 larries a day on a row of ovens 560 feet long, and assume that each larry would run to the farther end, the whole distance traveled in the thirty double trips during ten hours, it amounts to only 6.3 miles. Hence a speed of one mile per hour would be ample for doing the work at these ovens. Considering that three or four larries can be run out at the same time, **we may say that a speed of** from two to four miles per hour would be sufficient even for the largest coke works. With a speed **not** exceeding four miles, the rope would last about two years. There is a wide field here for inventive spirits to improve the grips for this purpose, and in fact there are already many patents in existence for an apparatus for gripping the rope **of passenger cable** roads which are in principle the same as the above-suggested **arrangement.** It is true, this requires double the length of rope that the Leisenring works have now in use, but the wages of two months for an extra engineer would pay for the additional **rope.**

Though the arrangements **described in the last chapter are** above ground, **they** are of the same **nature as** underground work, and **bear so intimate** a relation **with coal** mines that they could properly be placed under the same heading.

From the description given in the foregoing pages of the various methods of wire rope haulage in operation at so many coal mines, there can no longer be any doubt of **the** practicability of this method of transportation under **any circumstances.** The only question to **the practical man** will yet be: Is the **conveyance** of coal by wire **rope as economical as it is efficient?** This question can be answered in the **affirmative.** **Comparing the transportation by locomotives or by mules (the** only **two methods which can come into** consideration), we find that both are considerably **more expensive than haulage** by wire rope.

The possible application **of locomotives is** very limited, as they **can only be** used on certain grades **which are rarely met** with in mines. A great objection to **locomotives is also the inevitable smoke and foul air** produced by them. **There are mines** which **on** account of this inconvenience have abandoned **the use of locomotives.** It was also found that in spite of favorable gradients the expenses **for hauling** by locomotives were greater than they were formerly, when the whole haulage was done **by mules.**

Mules or horses are used in all mines for **drawing pit-cars from the working** rooms to the partings, or the reverse. **For very short distances** and on **not too** excessive grades they will probably **always be a convenient and** economical **motive power.** There **are, however, still many mines in which** mules **are** employed **exclusively, even where the coal** must be hauled long distances. **In** such cases the owners will always find it **to their advantage to** replace the mules **by a wire rope** arrangement.

Long experience has established **that the best** results of animal traction are **obtained on** grades of 1 foot in 130 feet in favor of **the** load, and at a traveling **speed of $2\frac{1}{2}$ miles per hour. The pulling** force of a mule walking at this speed **and working ten hours per day is 75 to 80 pounds. Within** small limits of these normal conditions the load to be pulled and the speed with which the animal travels can be varied in inverse proportion without changing appreciably the net effect of the performed labor, but the favorable results decrease rapidly with higher speeds, steeper grades or grades against the load.

Careful investigations made by the author in regard to the cost of a wire rope plane and the running expenses, compared with the cost of mules, the expenses of their maintenance and drivers' wages, have given the following result: On roads shorter than one-quarter of a mile, and on a grade in favor of the load of 1 foot in 130 feet up to 1 foot in 70 feet, there will be a difference of $\frac{43}{100}$ of a cent per ton in favor of mule haulage. If, however, the grade is against the load, the haulage by mules costs from 60 to 80 per cent. more than that by wire rope. If the road has a length of one-half mile the cost of mule haulage is 25 per cent. more than wire rope haulage on a grade of 1 in 130 in favor of the load, and 200 per cent. more if the same grade is against the load. The difference increases in the same proportion for longer roads; for instance, to convey coal a distance of $1\frac{1}{2}$ miles it will cost three times as much by mules as by wire rope on a favorable grade for the former, and more than five times as much on an unfavorable grade.

The average cost of wire rope haulage on an undulating grade is 2.2 cents per ton per mile, and that of mule haulage 7.6 cents.

In the majority of mines the grades are more unfavorable for mule haulage than was assumed in the calculation of this comparative cost, and the difference will be still more in favor of hauling by wire rope.

All the owners of the previously-described mines agree that without the rope machinery they would have to close the mines, as it would be impossible to convey the coal economically enough by any other method.

V. WIRE ROPE TRAMWAYS.

The methods of conveying coal and other mining products on a suspended rope tramway belong exclusively to overground haulage, and find especial application in places where a mine is located on one side of a river or deep ravine and the loading station on the other. A wire rope suspended between the two stations forms the track on which material in properly constructed "carriages" or "buggies" can be transported as quickly and safely as over the solid ground. It saves the construction of a bridge or costly trestlework, and is practical for a distance of 2000 feet without an intermediate support.

There are two distinct classes of rope tramways:

(a.) The rope is stationary, forming the track on which a bucket holding the material moves forward and backward, pulled by a smaller endless wire rope.

(b.) The rope is movable, forming itself an endless line, which serves at the same time as supporting-track and as pulling-rope.

Of these two the first method has found more general application, and is especially adapted for long spans, steep inclinations and heavy loads. The second method is used for long distances, divided into short spans, and is only applicable for light loads which are to be delivered at regular intervals.

Rope tramways of this kind are constructed in great numbers in all parts of this country, and serve a variety of purposes. They convey mining products across rivers, or stones from the depth of quarries to the banks, and in manufacturing establishments are frequently used to distribute material taken from boats or cars to certain storing places in the yards.

A few examples will give a clearer understanding of their construction and method of operation.

Fig. 88, Plate 22, illustrates the tramway used by Mr. Stanley Loomis, at Logansport, Pa. The rope is suspended in a single span of 1400 feet across the Allegheny river, and serves to convey coal and limestone from the west shore to the Allegheny Valley R. R. on the east shore. On the west side there is a high bluff, while the opposite side is low, making a difference of 190 feet in the height of the two points of suspension. The rope is of cast-steel, of 2 inch diameter, and contains, around a hemp centre, six strands of 19 wires each, with a total ultimate strength of 100 tons. A "carriage" constructed with three wheels (*Fig. 91*), with an iron bucket suspended to it, is pulled across by an endless steel wire rope of ¾-inch diameter, driven by an engine situated on the bluff of the west shore. The pulling-rope passes at each end around a single-grooved rubber-lined pulley of 3-foot diameter, the large span producing sufficient tension to prevent its slipping. Motion is given to one of the pulleys by a belt from the engine shaft. The time for a single trip is half a minute, while a round trip, including loading and unloading, can be made about every four minutes. The weight of the carriage and empty bucket is 1½ tons, and the bucket contains 3½ tons of material, making the total weight of the passing load 5 tons. This, together with the weight of the rope, produces a strain of 28 tons, or the three and six-tenth part of its breaking strength. Besides this direct strain there are local strains in the wires, arising from being bent around the wheels of the carriage. In a later chapter we shall show that these latter strains are considerable, and that the duration of a tramway rope depends largely upon the proper construction of the carriage.

The towers at either end, on which the rope is supported, can be constructed of a wooden frame-work, as shown in *Fig. 89,* and the connection with the cable is best made with a movable link (*Fig. 90*). This is preferable to passing the rope over the top of the timbers, because the great deflection under the passing load has a tendency to break the wires where they bend over the edge of the wood. As every rope stretches, it is advisable to provide it with an adjustment, consisting either of a turn-buckle to which the rope can be attached by means of a wrought-iron open socket, or of a screw-stirrup and cast-iron socket as shown in *Fig. 90.* Mr. Loomis states it as his experience that smooth cast-iron carriage wheels were less injurious to the wear of the rope than wooden wheels

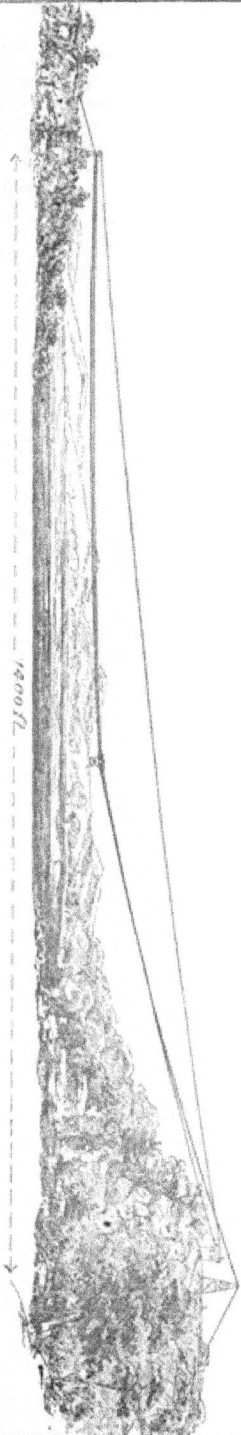

Wire Rope Tramway of Stanley Loomis
at Lansingburg, Pa.

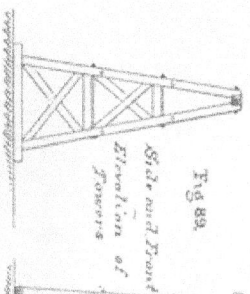

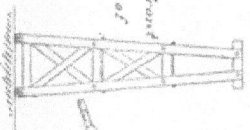

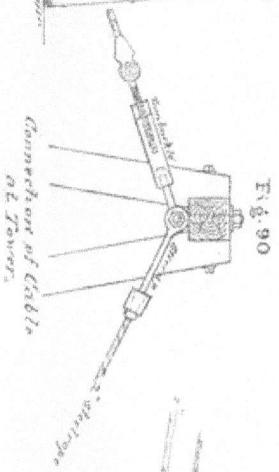

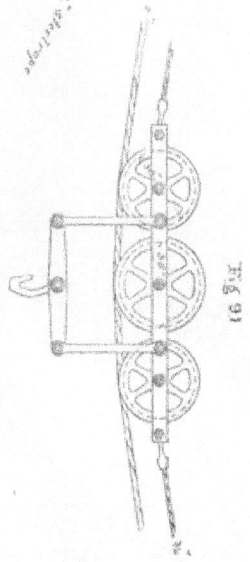

1400 ft.

Fig. 89.
Side and Front
Elevation of
Towers.

Fig. 90.
Connection of Cable
at Tower.

Fig. 91.
The Carriage.

Pl. 23.

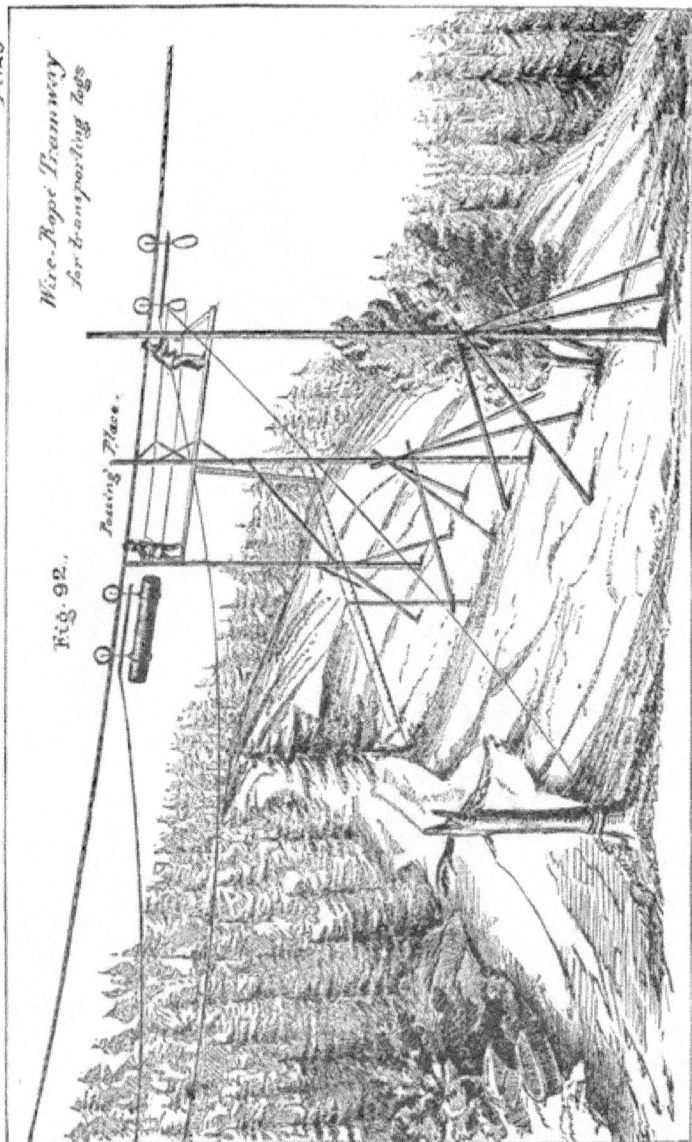

Pl. 23.

Wire-Rope Tramway
for transporting Logs

Fig. 92.

Passing Place.

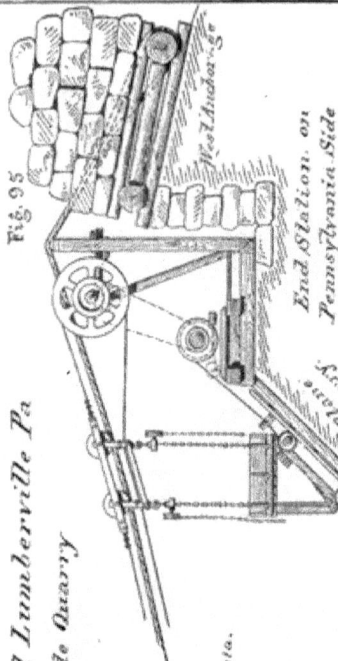

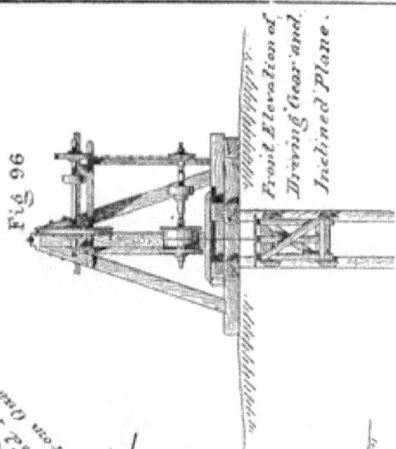

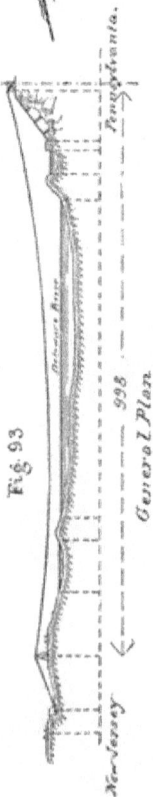

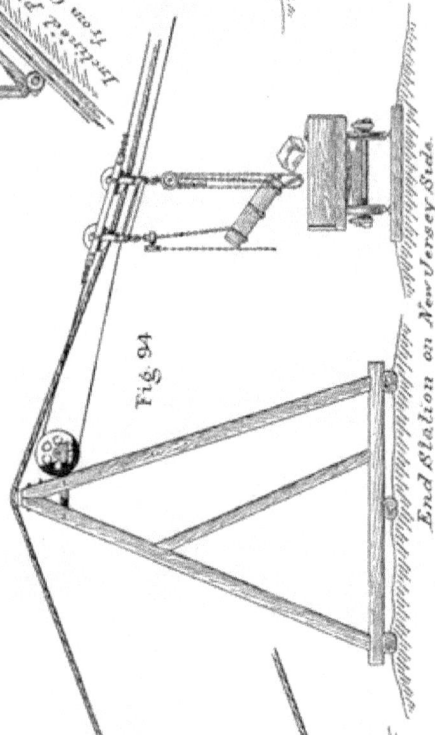

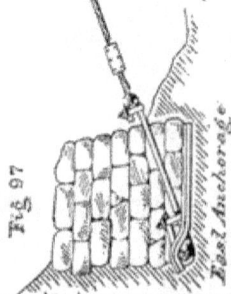

Pl. 24.

Wire Rope Tramway at Lumberville Pa

Kemble's Granite Quarry

Fig 95

West Anchorage

End Station on
Pennsylvania Side

Fig 96

Front Elevation of
Driving Gear and
Inclined Plane.

Incline from Quarry.

Incline from Quarry.

Fig 93

Delaware River

Pennsylvania

998

New Jersey

General Plan

Fig 94

End Station on New Jersey Side

Fig 97

East Anchorage

or iron ones lined with wood. As a rule, however, the wires are not worn very much, and they break not because their section is reduced, but because they are too often strained beyond their limit of elasticity by the continuous bending.

A type of wire rope tramway, which is frequently used in Switzerland for transporting logs from the top of a mountain to the valley, is represented in *Fig. 92, Plate 23*. The logs are suspended to a simple carriage of two wheels, and the rope has enough inclination to allow the loaded carriage to run down hill by its gravity, pulling up at the same time the empty carriage. For this purpose a small wire rope, passing around a drum or wheel on top of the hill, is connected to each carriage much in the same way as the cars on a gravity plane—this kind of tramway being in reality nothing else than a self-acting inclined plane suspended in the air. At the meeting place of the two carriages a light scaffold is erected for a couple of men to stand on, whose duty it is to lift the empty carriage off the rope and put it on again on the upper side of the descending loaded one. Of course, in stretching two ropes side by side the services of these men could be dispensed with and the travel of the two carriages be made automatic. A simple calculation of the comparative cost will in each case decide which method will be the most economical.

Another tramway of large dimensions is illustrated in *Plate 24, Figs. 93–97*. It is situated at Lumberville, Pa., and serves to convey granite blocks, broken from W. H. Kemble's quarry, across the Delaware river to the Belvidere Division of the Pennsylvania R. R. The distance from the Pennsylvania shore to the Jersey shore is 998 feet, which has been spanned with a 1⅜-inch steel rope, supported on each side on wooden towers, the ends of the rope being securely anchored.

The stone blocks are brought from the depth of the quarry on an inclined plane, operated by the same engine which drives the pulling-rope. A special car runs on the incline, taking a wooden box filled with paving blocks up or an empty one down. Each box is provided with a short chain, and when the car of the incline arrives on top, under the rope carriage, this chain is connected with the carriage, and the latter, with the box suspended to it, is now ready to be pulled across the river. One end of the chain passes over a small pulley, and when the carriage has reached the other shore a pull on this chain will easily tip the box and empty its contents into a railroad car. The pulling-rope is of ⅞-inch diameter, and motion is given to it by a grip-wheel on the Pennsylvania side, while the return-wheel on the opposite side consists simply of a grooved sheave. *Figs. 95* and *97* show different methods of anchoring the cable. This tramway, with all machinery and appliances, was planned and built by A. J. B. Berger.

In the Pennsylvania slate region of the Blue Ridge mountains the wire rope tramways have proved to be of valuable service, and are constructed in great numbers. The general type of them we represent on *Plate 25, Figs. 98–102*, taken from the Old Bangor quarry. The quarry is very extensive, and four or five ropes, at short distances from each other, are stretched from the height of one bank to a point on the opposite bank low enough to give the rope sufficient inclination for the carriage or buggy to descend by force of gravity. The following is the method of operation:

The empty buggy, held in position at point (*A*) by a frame (*S*) (*Fig. 99*), is made free by pulling at the rope (*J*) and placing the frame in the position (*S¹*). It descends by its own gravity, pulling the hoisting-rope (*H*) after it, which uncoils from the drum (*Q*). This drum runs loose on the shaft, and is provided with a brake to enable the engineer to regulate the downward motion of the buggy. At a certain point of the rope, above a place in the quarry from which the stones are to be taken, an iron block (*W*) called "the jack" is secured to the rope, stopping the buggy from going farther (*Fig. 100*). The hoisting-pulley (*L*) sinks vertically by its own weight, while the hoisting-rope still uncoils from the drum and follows the pulley. To prevent it from being dragged through the mud or over the stones, it is supported by a roller (*P*) on the little carriage (*R*). A light hemp rope connects this carriage with a wooden counterweight (*C*), which slides on the back cable and can stop the descent of the carriage at any desired point. When the buggy returns it pushes this supporting-carriage back, while the counter-weight (*C*) sinks toward the anchorage; and when the buggy goes out again the little carriage follows by its own weight until stopped, because the counter-weight has reached the top of the derrick. After the slate, either in a single block or in smaller pieces in a box, has been attached to the hook of the hoisting-block, the engineer puts the drum (*Q*) in gear and hoists until the pulley (*L*) reaches the carriage in the position shown in *Fig. 100*. In continuing to hoist, the pulling-rope causes the buggy with its load to ascend the rope towards the derrick until reaching point (*A*), where it is held in position by the frame (*S*). As soon as the strain is taken away from the hoisting-rope the drum is put out of gear and the load immediately sinks to the ground. After it has been taken off, the drum is put in gear, the block hoisted, the frame (*S*) raised, then the drum again put out of gear, and the buggy starts out on another trip, all operations being repeated.

The stop-block (*W*) slides down the rope by its own weight, and is held in position or can be moved upward by a ½-inch wire rope, supported on small rollers in the buggy and little carriage, running over the derrick to an extra drum, which also can be placed in and out of gear according to necessity. The buggy in its present construction is the invention of Charles Shuman, and it performs its work with perfect satisfaction.

It is evident, however, that for the success of this arrangement two things are necessary: first, the friction resisting the downward motion of the buggy must be smaller than the friction resisting the vertical descent of the empty hoisting-block; and, secondly, the friction resisting the upward motion of the buggy must be larger than the friction resisting the ascent of the loaded hoisting-block. If these two conditions do not exist, in the first case the pulley would run down before the buggy could descend to its desired place, and in the second case the buggy would move up before the stone had been hoisted, making successful work impossible. The necessary favorable conditions depend upon the inclination of the rope, the diameter of the carriage wheels and the proportion

Wirerope Tramway
at
Old Bangor Quarry, Pa.

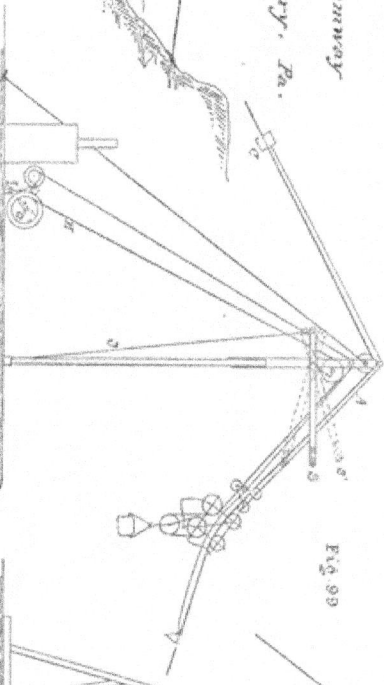

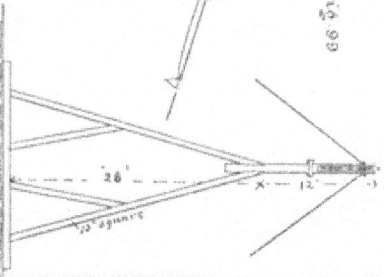

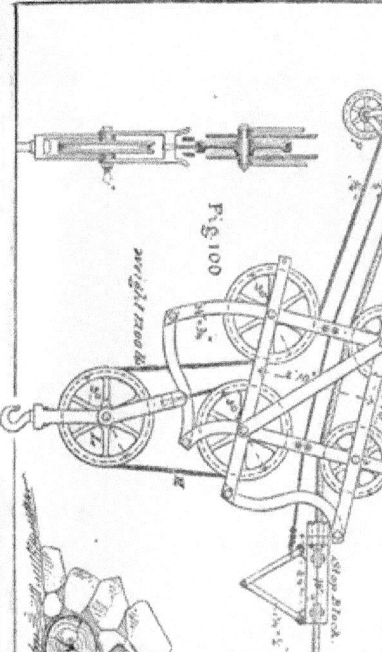

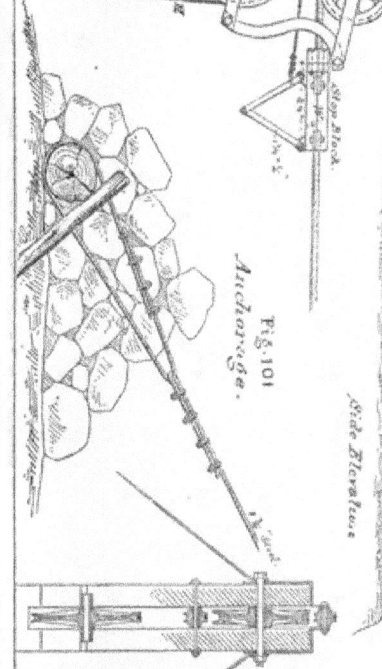

Fig. 98

Fig. 99

Fig. 100

The "Buggy"

Fig. 101
Anchorage.

Fig. 102.

Side Elevation

Front Elevation

Pl. 25.

Pl. 26.

Wire Rope Tramway at
Burden Iron Works, Troy N.Y.

Fig.103.

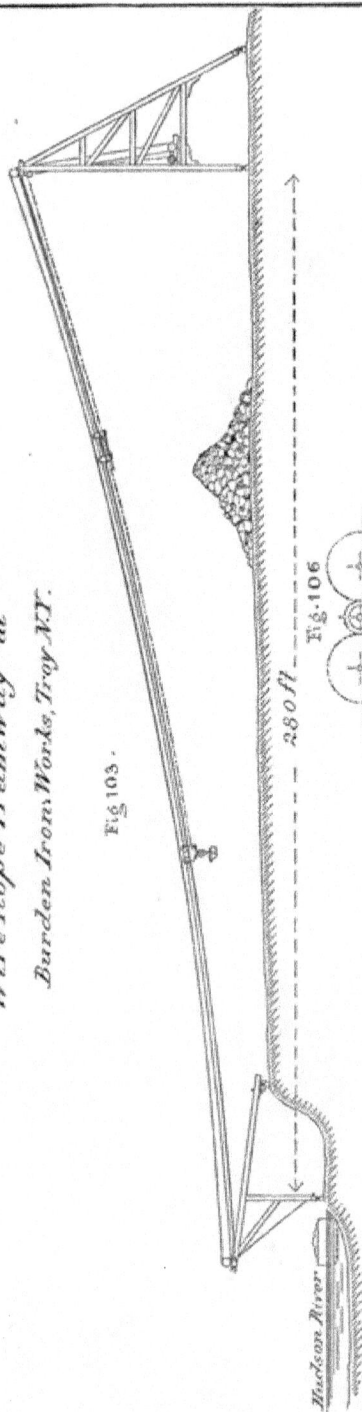

Hudson River

280 ft

Fig.104

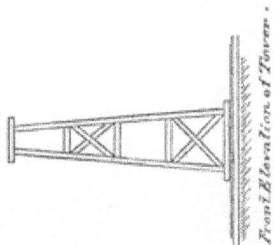

Front Elevation of Tower.

Fig.106

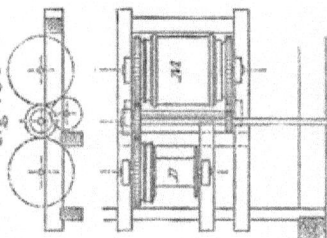

Hoisting Gear and Brakes.

Fig.105.

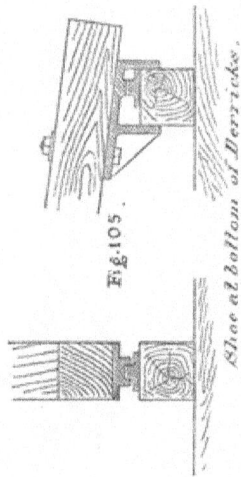

Shoe at Bottom of Derricks.

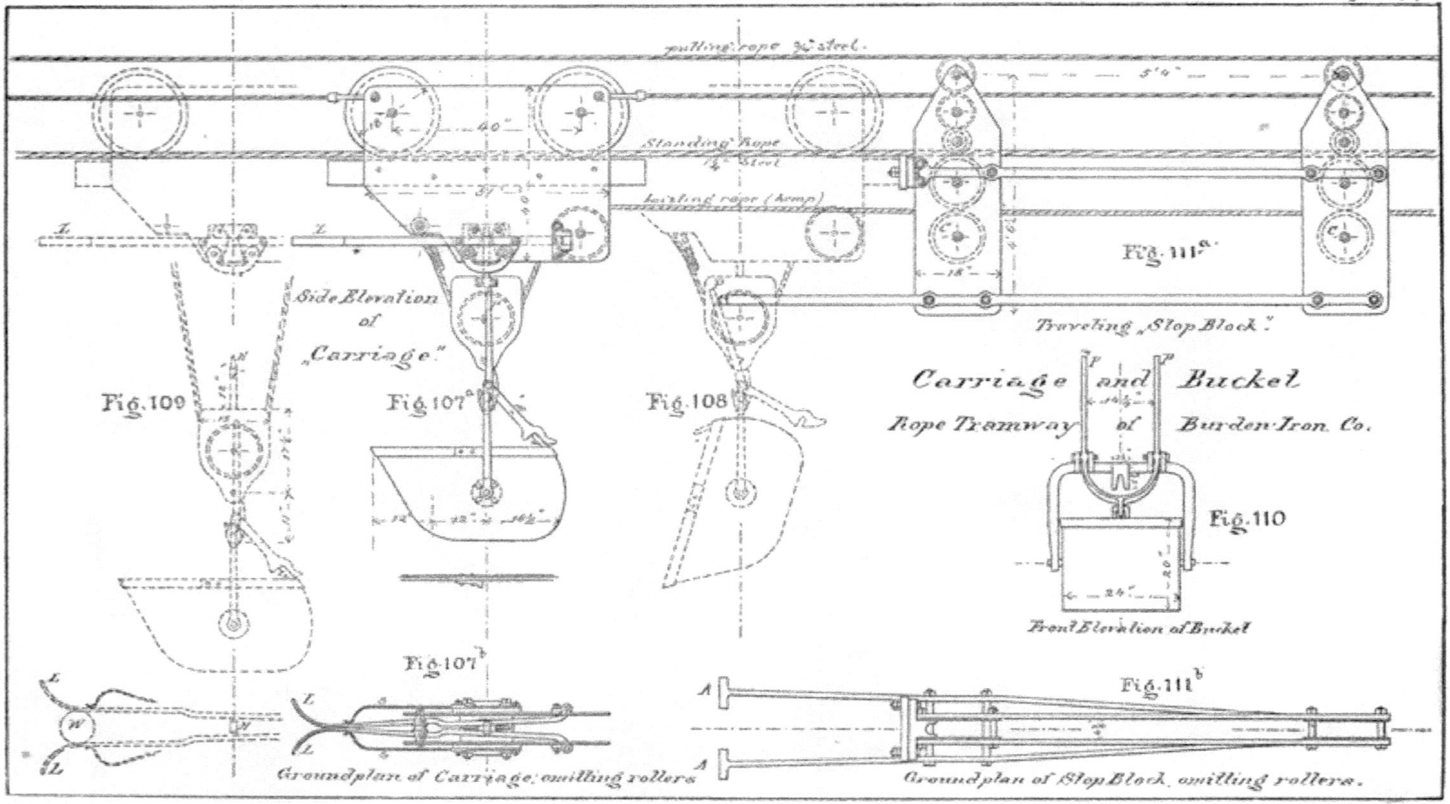
pulling rope ¾" steel.
Standing Rope 1¼" Steel
hoisting rope (hemp)
40"
37"
5'4"
18"
Side Elevation of „Carriage."
Traveling „Stop Block."
Carriage and Bucket
Rope Tramway of Burden Iron Co.
Front Elevation of Bucket
24"
20"
Groundplan of Carriage; omitting rollers
Groundplan of Stop Block, omitting rollers.

Fig. 109

Fig. 107ᵃ

Fig. 108

Fig. 111ᵃ

Fig. 110

Fig. 107ᵇ

Fig. 111ᵇ

between the weights of the carriage and the empty and loaded hoisting-block. The buggy, with hoisting-pulleys, weighs 1200 pounds, and the load of stone varies from half a ton to two tons. The cost for buggy, jack and supporting-carriage is about $150. The hoisting engine has an upright cylinder 8x12 inches, a gearing of a 4-foot spur-wheel with 8-inch pinion, and a drum of 32-inch diameter. The span of the tramway is about 450 feet, but the pulling-rope is 1000 feet long, of which 900 feet can be coiled on the drum in one minute. The duration of the hoisting-rope is about one year, while the standing rope lasts generally from two to four years. Both are steel ropes, with 19 wires to the strand, the first of ¾-inch, the second of 1¾-inch diameter. *Fig. 102* shows the top of the derrick, with the rollers over which the different ropes pass, and *Fig. 101* illustrates the method of anchoring the back cable.

The wire rope tramways of the Burden Iron Co. (*Plates 26, 27, Figs. 103–111*) are used to transport coal and iron ore from the boat landing to certain storing places in the yard. Differing from the systems so far described, these tramways have the peculiarity that they can be shifted sideways from one place to another, so that it would be possible with one rope to distribute material over the whole yard, though in reality the company employs five ropes in a width of about 500 feet. For this purpose the towers or derricks (*Figs. 103–105*) are provided with cast-iron shoes at the foot of each post, and rest on iron rails secured to heavy timbers, which run along the whole width of the yard. The shoes are shaped in such a way that it is easy to move the towers in the direction of the rail, and at the same time to prevent them from being lifted up. The rear post of the tower serves, therefore, also, as anchorage for the cable in case the strain in the same should produce a greater upsetting movement in the tower than the latter's own weight could resist. An ingenious construction of the carriage and stop-block makes the lowering, hoisting and emptying of the bucket automatic. *Figs. 107–109* show the bucket in three different positions—when traveling along, when emptying its contents, and when being lowered to the boat. It will be noticed that when the carriage arrives at the stop-block (*Figs. 111ᵃ, 111ᵇ*), a bell-crank-shaped lever (*P*) of the bucket strikes against the arm (*A*) of the stop-block, unhooking the bucket, which, being heavier in front when loaded, tilts and empties the material (*Fig. 108*). When pulled away from the stop-block, the bucket, being balanced when empty, rights itself again. Two bars (*L*) secured to the carriage frame and pressed together by the springs (*S*) support the hook (*H*) of the block to which the bucket is suspended. At the outer derrick there is a pin (*W*), placed in such a position that it presses the two arms (*L*) apart, releasing the hook (*H*) and enabling the bucket to be lowered. The hoisting-rope is of hemp, the pulling-rope of the carriage of ¾-inch steel. The latter passes around a pulley at the outer derrick, and at the inner derrick is guided over two sheaves down to drum (*M*), around which it takes several turns to insure friction (*Figs. 103, 106*). The stop-block is provided, besides its running wheels, with several rollers for the support of the pulling and hoisting rope. For changing its position a small rope is attached to it, reaching to the ground, so that the block can easily be moved by hand.

Another hoisting and conveying apparatus, constructed with the object of stopping the traveling carriage at any desired point of the tramway rope, has been patented by M. W. Locke. It consists (*Fig. 111ᵉ*) of a two-wheeled carriage, to

Fig. III ᵉ

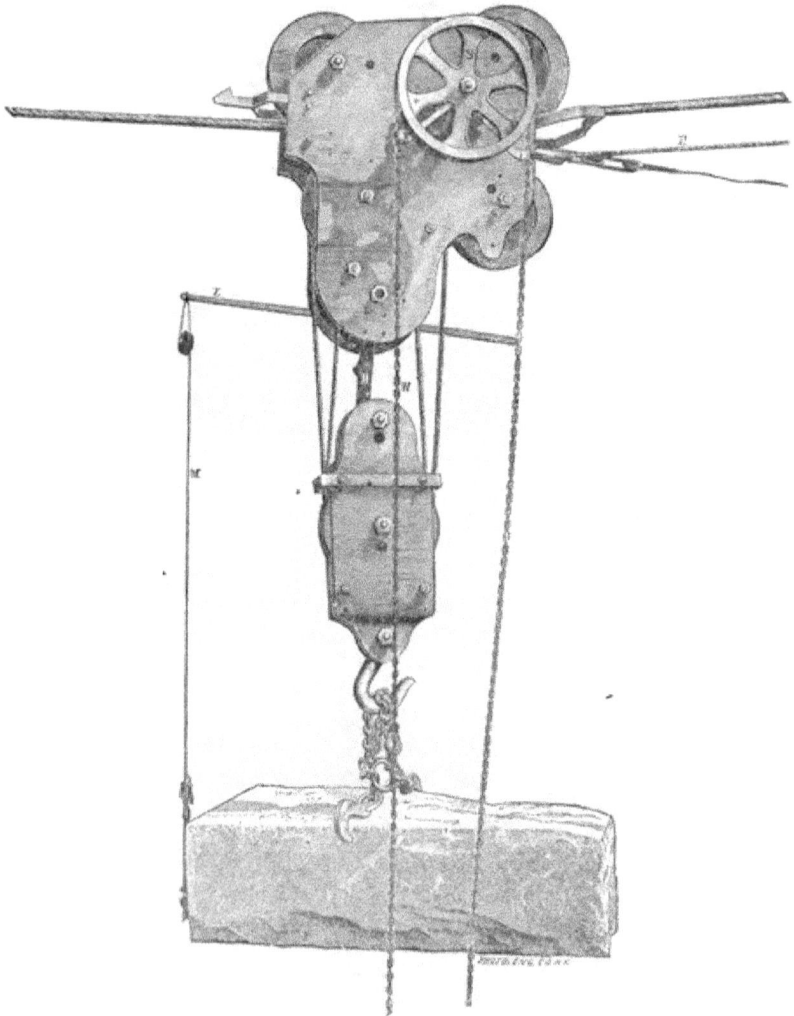

which the hoisting-block is attached by a hook (H). A short piece of rope connecting the carriage frame with the hoisting-rope prevents the block from being hoisted higher than the hook, because as soon as this short piece of rope becomes tight the pull of the hoisting-rope is transferred to the carriage,

Pl. 28.

Wire Rope Tramway with elevated Track,

Berger's Patent.

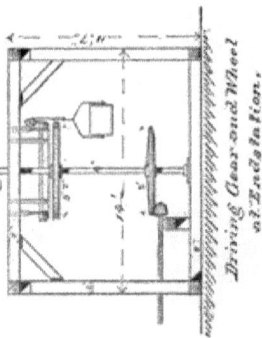

Fig. 115ᵃ

Driving Gear and Wheel at Endstation.

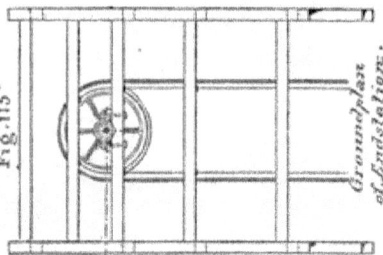

Fig. 115ᵇ

Groundplan of Endstation.

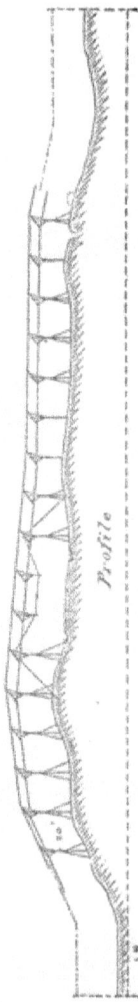

Profile

Plan of Tramway from Guiterman's Mines to Red Lion Station, Berks Co. Pa.

Fig. 112

Fig. 113

Cross Section of Tramway.

Fig. 116

Angle Station at L.

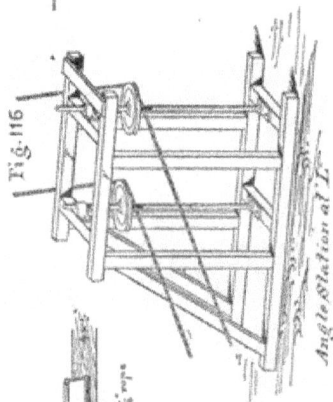

Fig. 114ᵇ

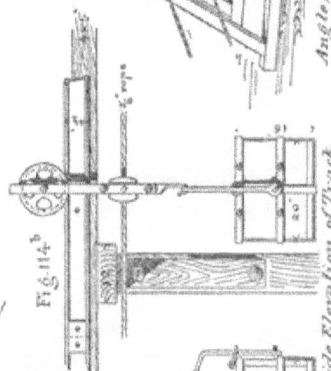

Fig. 114ᵃ

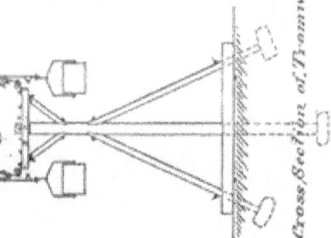

Front and Side Elevation of Track and Bucket.

moving the same in place of hoisting the block. At the point where the load is intended to be lowered a man pulls the chain (W), turning the chain-wheel (S), which works a screw and tightens a grip within the carriage frame of similar construction to that shown in *Fig. 76*, *Plate 18*. The hook (H) is connected with a long lever (L), and in pulling the rope (M) the block is made free from the hook and can be lowered. The apparatus is adapted only to places where the tramway rope has sufficient inclination to allow the carriage to descend by force of gravity, and where the space under the rope is free of obstructions which would interfere with the long ropes and chains hanging from the apparatus and necessary to grip the carriage and to unhook the hoisting-block.

A different class of tramway is that employing an elevated iron track in place of a suspended wire rope, but otherwise worked in the same way. There are several systems of this kind in existence, one of the best being Berger's patent, illustrated in *Figs. 112–116*, *Plate 28*. These tramways are especially adapted for heavy work and a continuous delivery of some mining product. Berger's system consists of an elevated track resting on wooden posts placed twenty feet apart. A number of buggies with a single wheel are placed at regular distances on this track; all of them are connected to and set in motion by an endless wire rope, which passes at each end of the tramway around a grip-wheel or grooved rubber-lined wheel, and is driven by an engine (*Figs. 115ᵃ, 115ᵇ*). The speed of the rope is about 150 feet per minute, which makes it possible to fill or empty the buckets at the ends while they pass around the end curve, and without taking them off. The rail is $4\frac{1}{2}$ inches high, having the section of an **I**, and is screwed to a longitudinal 4x5-inch timber (*Figs. 114ᵃ, 114ᵇ*). As a special advantage of this system it must be remarked that it is easy to go around any curve by simply placing a pair of wheels in the angle as shown in *Fig. 116*. The track has been omitted in this sketch to avoid the confusion of too many lines; it runs parallel to and directly over the rope suspended from the timbers (F), similarly as shown in *Fig. 115ᵃ* at the end station. The pulling-rope has a diameter of $\frac{3}{8}$-inch, and little power is required to move the buckets, as the rolling friction on the smooth rail is very small and only the axle friction has to be overcome. On steep inclinations of course more force, as well as a larger rope, is necessary. This is also the case if the number of buggies is increased. Making the pulling-rope and driving power large enough, there is scarcely any limit to the quantity of material which can be transported, as the buckets may be placed only ten feet apart.

(b.) TRAMWAYS WITH MOVABLE ROPES.

The first tramways of this kind were constructed in this country by John A. Roebling for the purpose of carrying the wire from one shore of a river to the other in making cables for his suspension bridges. He gave it simply the name "traveling rope" or "working rope." It consisted of a $\frac{5}{8}$ to $\frac{3}{4}$-inch wire rope, stretched from shore to shore, and spliced endless after passing it at each side around horizontal wheels. Motion was given to the driving-wheel by horse or steam power. A light wheel was securely fastened to this rope, carrying the wire across the river and returning empty.

This same principle has later been applied for the transportation of mining products over mountains, valleys or rivers. The rope for this purpose is supported every 100–150 feet, and a series of buckets are attached to it. A number of patents are in existence for the details of this system, among which we may mention Hodgson's patent. His method of attaching the bucket consists of a ∧-shaped clutch, which simply rests on the rope, being held by friction; it easily runs over the supporting-rollers. At the end stations the clutch passes from the rope to a separate rail, for which purpose it is provided with two small wheels, so that it can easily be pushed by the men in attendance around the curve to the opposite side, where it clutches the rope again. This passage of the buggies around the end rail is made use of to fill or empty them, according to necessity. As far as we know, these tramways have only been attempted in a straight line, but there is no particular objection to using them also in curves. At each turning-point, of course, it is necessary to place a wheel and rail, the same as at the end stations; also a man for pushing the buckets over the rail. An objection to this system lies in the fact that at inclinations exceeding 25 per cent. the clutch commences to slip, especially near the supporting-rollers, where occasionally two buckets collide and are thrown off the rope. For light loads and in localities where the erection of many posts is objectionable, these tramways are preferable to those with fixed rails, and do very good service.

INCLINED PLANES.

Several systems already described in the underground haulage may be termed "inclined planes," but this name proper is generally only given to the large overground planes of the different coal railroads, on which whole trains are alternately raised or lowered. They are nearly all constructed after the principle of the endless rope system, but they differ somewhat in details and are especially remarkable for their large dimensions, so that it will be of interest to mention some of the principal ones.

Fig. 117, *Plate 29*, represents the inclined plane of the Lehigh Coal and Navigation Co., at Solomon's Gap, near Wilkesbarre, Pa. There are three inclines above each other, with a combined length of 14,800 feet, and 6000 tons are raised per day to a height of 1640 feet, at the price of one cent per ton per mile. The plane is worked by an endless-rope system, composed of a $2\frac{1}{2}$-inch iron main rope, wound three times around two drums of 19-foot 9-inch diameter, and of a $1\frac{1}{2}$-inch following or tail rope, passing around a return-wheel of 12-foot diameter, which is placed at the foot of the plane on a movable carriage, with a weight attached to it to keep the rope constantly in equal tension. The drums reverse alternately for raising the load, consisting of 24 cars, either on one track or the other.

The driving engine is of 800-horse power, and consists of two vertical cylinders of 26-inch diameter by 40-inch stroke, and a gearing with a 6-foot pinion and 19-foot 9-inch spur-wheels.

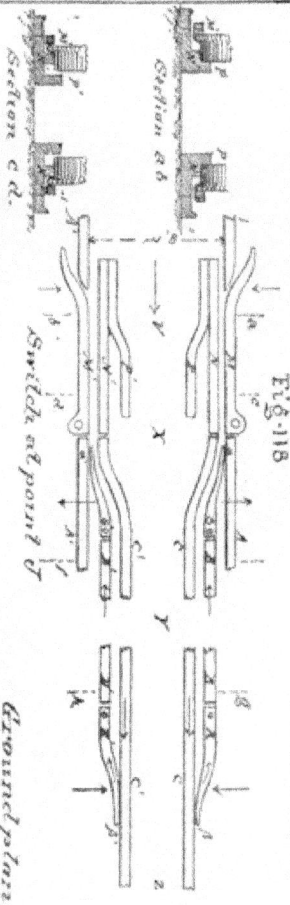

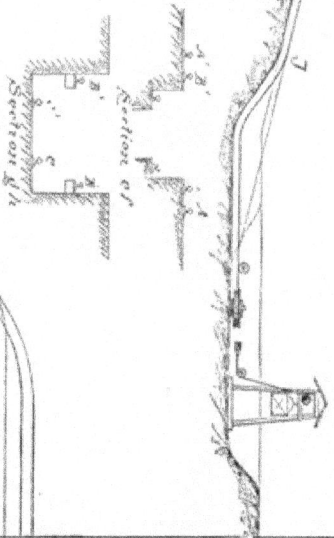

Inclined Plane at Solomons Gap near Wilkesbarre.

Fig. 117ᵃ

Fig. 118

Fig. 117ᵇ

Section a b

Section c d

Switch at point J

Ground plan.

Elevation.

Section e f

Section g h

Pl. 29

Pl.30

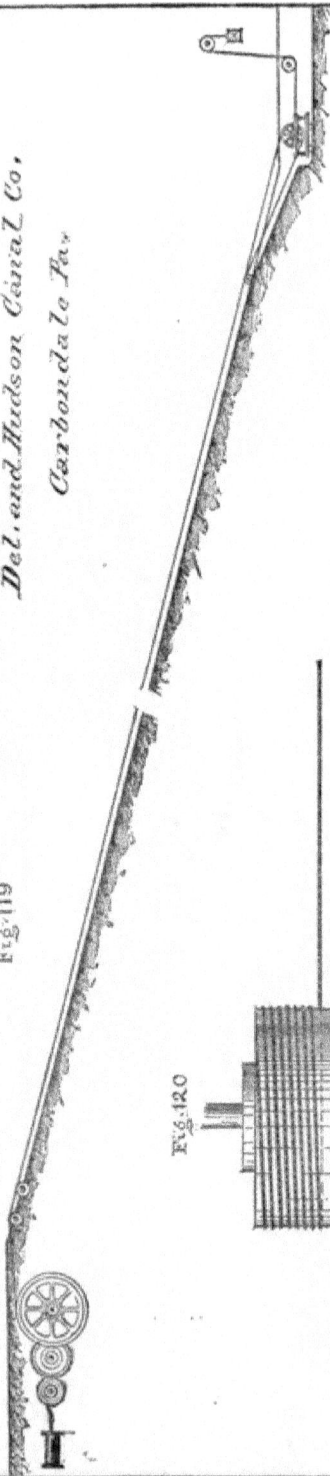

Inclined Plane
Del.and.Hudson Canal Co.
Carbondale Pa.

Fig.119ᵃ

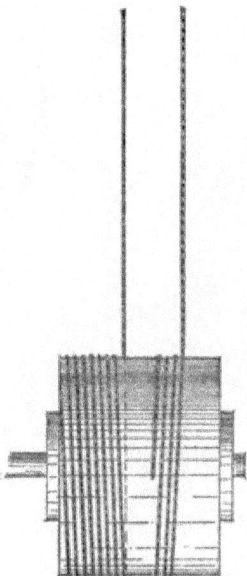

Fig.119ᵇ

Ground plan

Fig.120

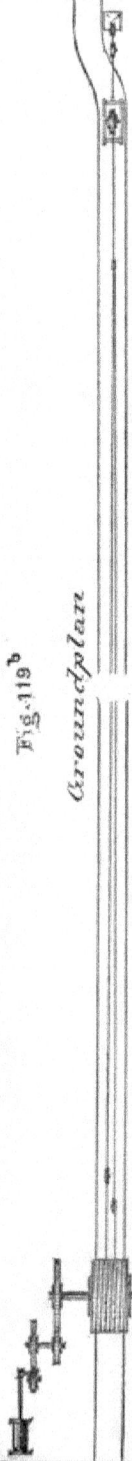

Each train is accompanied by a special car, called the "barney," which is attached to the lower end of the train, pushing the cars up the incline, while the other barney descends empty on the opposite track. The ropes are fastened to the barneys by means of sleeve sockets. In order to save an extra pair of rails for the barney and nevertheless to allow the other cars to pass it at the bottom of the incline, an ingenious switch has been contrived and placed at point (J) (Fig. 117*). By means of this switch the barney descends into a pit from track $(A\ A^1)$ to track $(C\ C^1)$, while the coal cars continue on track $(A\ A^1)$ in the same level (Fig. 118). To effect this the wheels of the barney are loose on the axles and movable on them, the flanges are outside, and it can therefore readily be understood that the wheels, when touching the tongue (M), are pushed inward so as to run first on the rails (N), and afterwards are still more pushed together by the rail (S) until they run ultimately on the rails $(C\ C^1)$. The first motion of the wheels is limited by the projecting rail (P), and the second by a shoulder on the axle. Both rails (M) and (S) are movable around pivots, but held in position by springs pressing in the direction of the arrows. On the return trip the reverse takes place. When reaching points $(\beta\ \beta^1)$ the wheels are pressed apart by the tongues $(R\ R^1)$ so as to force them to run on track $(B\ B^1)$, from where they pass over on $(N\ N^1)$, and by means of the rail (P) projecting over $(N\ N^1)$ are finally pushed on $(A\ A^1)$ again. For a still better understanding of Fig. 118 a few sections are drawn, showing the position of the rails and car wheels at different points of the switch

The plane has two tracks, on which the full trains ascend alternately, and at the top as well as at the bottom there are the necessary switches and side tracks for the arrangement of the train of cars (Fig. 117*).

Another famous series of inclined planes are those of the Delaware and Hudson Canal Co. between Carbondale and Honesdale. They cover a distance of 17 miles, there being 17 inclined planes, varying in length from 1000 to 1500 feet, and in inclination from 10 to 12 feet in one hundred. Fig. 119, Plate 30, represents one of these planes in elevation and ground plan. Some are worked upon the main and tail rope system, the ends of the rope being fastened to the drum as in Fig. 120; others upon the principle of the endless rope, the rope not being fastened to the drum, but only wound on it with three or four turns to get enough friction to prevent its slipping; the latter method is used only for lowering empty cars to the repair shops. In both cases the return-wheel is placed on a carriage with a weight attached to it to keep the rope in constant tension. A train attached to the rope generally consists of 5 cars, each containing a load of 5 tons of coal. The train is taken up the plane with a velocity of 18 to 19 miles per hour. Each incline is worked by an engine of 250-horse power, of the type shown in Fig. 119, with two horizontal cylinders. The heaviest ropes employed on the ascending inclines have a diameter of $1\frac{1}{4}$ inches, and the smallest 1 inch. Each plane has only a single track, but there are two separate series of planes—one for the east-bound full trains, the other for the west-bound empty trains. On all the inclines for the ascending trains, safety arrangements are placed at certain distances to guard against accidents in the case of a rupture of the rope. These planes were first constructed in 1827, before the introduction of the locomotive, and are at the present day still found to be the cheapest method of transporting coal between the two points named.

There are many other inclined planes of equally great dimensions, of which we mention yet those of the Pennsylvania Coal Co., the inclined plane at Mahanoy, Pa., and one at the former Mine Hill and Schuylkill Haven R. R. (now Philadelphia and Reading), of which the plan of the machinery is illustrated in *Fig. 121, Plate 31.*

As a matter of historical interest we mention the old Portage railway incline (*Fig. 122, Plate 32*), which formerly served to take canal boats of the Pennsylvania canal from Hollidaysburg, at the eastern base of the Alleghenies, over the mountains to Johnstown, at the western base. It was originally built by Bertrand, one of Napoleon's old generals, and was operated with hemp rope. In 1840 it was reconstructed for the use of wire rope by John A. Roebling, and was for ten years in operation, but since 1850 it has been abandoned.

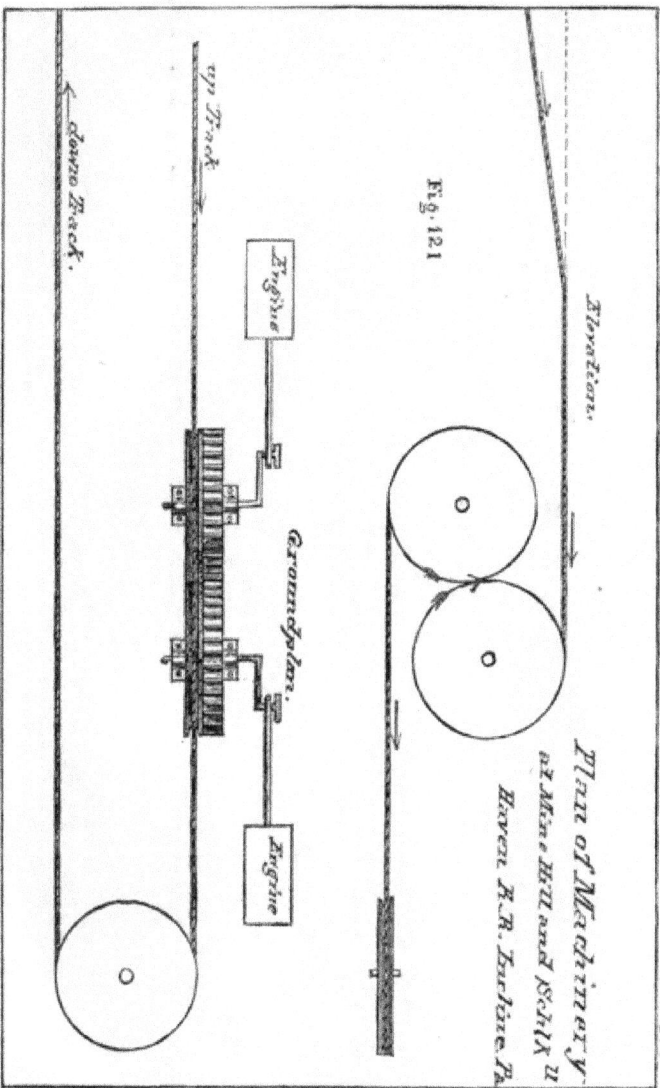

Fig. 121

Elevation.

Ground plan.

Engine

Engine

up Track

down Track

Plan of Machinery
at Mine Hill and Schll
Haven R.R. Incline Pa

Pl. 31.

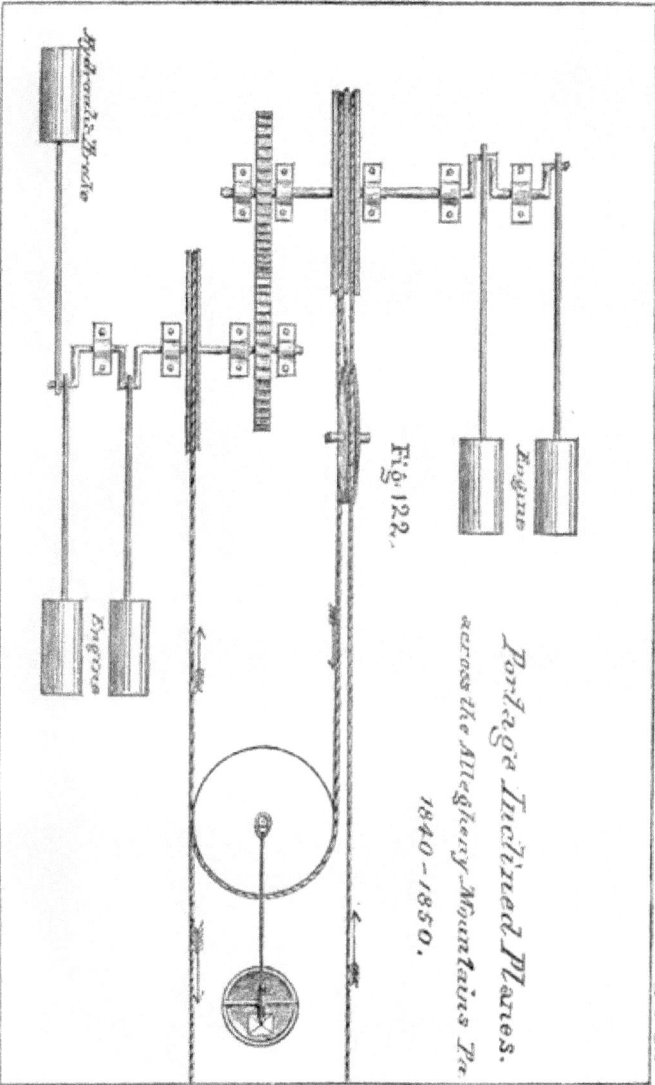

Fig. 122.

Portage Inclined Planes.
across the Allegheny Mountains Pa
1840-1850.

Pl. 32.

Wire ropes are usually made of 6 wire strands, laid around a hemp heart or centre. A greater or lesser number may be used, but it is seldom done. For special purposes a wire strand is sometimes substituted for the hemp centre; at times, also, a hemp centre is put in each of the strands. Each wire strand is composed of either 19 or 7 wires; any other number does not make a compact strand, and is therefore not advisable. Using either 19 or 7 wires, and 6 strands with a hemp heart, gives 114 and 42 wires respectively for the total number in the rope; its strength is therefore equal to the aggregate strength of the 114 or 42 wires, less 10 per cent., which loss is due to the twisting. The number of wires and the "lay" of the rope, whether long or short, have advantages and disadvantages.

The opinions of mine superintendents vary much as to which kind of rope is best. Special conditions govern almost every case. In the mines of the Monongahela region preference is mostly given to steel ropes, with 7 wires to the strand, and made with moderately long lay.

For general rules regarding the different kinds of wire ropes it may be said:

1. Ropes with 19 wires to the strand, being more pliable, are preferable in vertical hoisting and in cases where the rope is led around sharp curves, provided it does not drag over the ground and that friction is avoided as much as possible.

2. Ropes with 7 wires to the strand are stiffer and require larger drums or sheaves than those with 19 wires, but the thicker wire can stand considerable wear, and these ropes are therefore preferable on straight or nearly straight roads, and where the rope is exposed to much abrasion and other injuries.

3. Ropes with a long twist stretch little and glide easily over rollers; they are therefore well adapted for the tail rope system and wire rope tramway.

4. A short-twisted rope is very elastic, and consequently stretches considerably. On account of this property such a rope is to be recommended for all inclined planes, where the rope is occasionally exposed to sudden dangerous shocks which may prove fatal to the less elastic long-twisted rope. It is also easier to make a stronger and more durable splice in a short-twisted than in a long-twisted rope. For the endless rope system, where a good splice is a necessity but where at the same time a great stretch is inconvenient, a medium twist is therefore preferable.

63

5. In going around curves it is always better for the wear of the rope to lead it over one single sheave, provided this is made large enough, than over a number of smaller rollers.

6. In most cases a steel rope is to be recommended in preference to an iron rope. It is cheaper than an iron rope of equal strength; also much lighter, less bulky, more elastic, harder, and therefore more durable. On the other hand, sometimes its elasticity is inconvenient, causing the rope, when wound on a small drum, to uncoil and jump off after the strain has been released. Its hardness, though a good quality for the rope, is injurious to rollers and sheaves, wearing them out more rapidly than an iron rope.

This variety of qualities makes it possible to select in any case a wire rope most suitable for the desired purpose.

The durability of a rope depends principally on the diameter of the drum or sheave around which it is coiled. If an iron bar or single wire is bent, certain fibres are elongated, others contracted, producing a tensile or compressive strain equal to the force of a direct pull or pressure which would elongate or compress the fibres to the same extent. The quantity of this force, and hence the strain per unit of sectional area, depends upon the modulus of elasticity of the material, the thickness of the wire, and the proportion of the elongation to the original length. The smaller the drum, the sharper is the bend and the greater the strain; therefore in determining the size of a drum or sheave, the consideration is guiding that the strain produced by bending, combined with the direct pull of the working load, should not exceed a certain maximum. For this maximum we take the limit of elasticity of the material—the limit to which it can be strained a great many times without permanent injury. From the nature of the rope it follows that the size of the drum does not depend upon the diameter of the rope, but only upon the diameter of the wire of which it is made. It is true that in consequence of the twist a certain friction exists between the individual wires of the rope, but it is so small originally, and with a free application of oil is still more reduced, that it can safely be neglected; consequently the drums need not be larger than for a single wire.

The following table has been calculated under the assumption that the working load of the rope is one-fifth of its ultimate strength, and that the modulus and limit of elasticity, both for steel and iron, have an average value based on the latest researches.

Diameter of Rope.	SMALLEST DIAMETER IN FEET OF DRUM OR SHEAVE.			
	Steel Ropes.		Iron Ropes.	
	19 Wires to the Strand.	7 Wires to the Strand.	19 Wires to the Strand.	7 Wires to the Strand.
Inches.	Feet.	Feet.	Feet.	Feet.
2¼	8.6		13.0	
2	8.0		12.0	
1¾	7.2		9.5	
1⅝	6.3	.....	8.6	
1½	5.7	8.6	7.8	13.0
1⅜		8.0	7.6	12.0
1¼	5.0	7.2	6.7	10.8
1⅛	4.5	6.3	6.0	9.5
1	4.0	5.7	5.4	8.6
⅞	3.6	5.0	4.6	7.6
¾	3.0	4.5	4.0	6.7
11/16	.....	4.0	.....	6.0
⅝	2.3	3.6	3.4	5.4
9/16	1.7	3.0	2.6	4.6
½	1.5	2.6	2.3	4.0
7/16		.....		3.4
⅜		2.0		2.8
5/16		1.7		2.6

It appears from this table that, contrary to the ordinary belief, iron ropes require larger drums and sheaves than steel ropes. This is owing to the fact that iron wire, having about the same modulus of elasticity, possesses only an ultimate strength and a limit of elasticity of less than one-half that of steel wire. There are frequently practical reasons for choosing, in certain cases, larger drums than the sizes stated in the table; for instance, to avoid the recoiling and jumping off of the steel ropes after releasing the tension.

If the working load produces in the straight part of the rope less strain than one-fifth of its breaking strength, the drum diameters may be smaller without injury to the rope, but if the working load is greater than one-fifth of the rope's ultimate strength, the drums must be correspondingly larger if the strain shall not exceed the limit of elasticity.

In leading wire ropes around curves it is often impossible, for lack of space, to use a large sheave, and recourse must be had to a number of small rollers. With this arrangement many mistakes have been made, in consequence of which there has been a speedy wearing-out of the rope. The success in one case, the failure in another, and the varied opinions of practical men concerning the best methods, are a proof of this, and demonstrate the importance of the matter. Close observation and the comparison of many facts collected in the Monongahela coal regions seemed to indicate that similarly to the law governing the diameter of a single-wire sheave there would also be another law determining the number and position of the small rollers, so that no part of the rope would be strained beyond its limit of elasticity. Theoretical investigations corroborate this, and show that with the proper arrangement a rope can be taken around a curve by means of small rollers with the same safety as by means of one large sheave. This is of great advantage in practice, but only true when the rollers are correctly arranged. A general rule cannot be given for such an arrangement on account of differing circumstances. It is necessary to investigate each case separately and to go through the whole course of calculations, but the benefit derived from it in doubling or tripling the durability of a rope is well worth the trouble.

Attempts have been made to reach a similar result by special construction of the supporting-rollers, of which O. H. Jadwin's patent supporting arrangement is an example. It consists of two rollers placed at the ends of a short beam which is pivoted in the centre. While in regard to the easier curvature of the rope this arrangement only approaches the true necessity, it has certainly the advantage of always giving to the rope a support, and preventing the vertical vibrations. It is also claimed for it that a grip as used in the endless rope systems and cable railways would slide easily over the rollers without the necessity of lifting the rope above the same.

A theoretical investigation is of special importance for suspended tramway ropes. The rapid wearing-out of these ropes is in most cases due to a faulty construction of the carriage; generally the wheels are either too small or not enough in number. With the proper construction of a carriage, adapted to the conditions of the weights, the life of a rope can be considerably prolonged.

www.ingramcontent.com/pod-product-compliance
Lightning Source LLC
Chambersburg PA
CBHW021820190326
41518CB00007B/669